lamentals of Cellular and Molecular Biology

s: Sanaullah Sajid, Sajjad ur Rahman, Shahid Mahmood,

ı Bashir and Mudasser Habib

(Online): 978-981-5238-03-7

(Print): 978-981-5238-04-4

(Paperback): 978-981-5238-05-1

ublished in 2024.

Fundamentals of Cellul
Molecular Biolog

Edited By

Sanaullah Sajid
Institute of Microbiology
University of Agriculture
Faisalabad, Pakistan

Sajjad ur Rahman
Institute of Microbiology
University of Agriculture
Faisalabad, Pakistan

Shahid Mahmood
Institute of Microbiology
University of Agriculture
Faisalabad, Pakistan

Shayan Bashir
Premier Research, Ontario
Toronto, Canada

&

Mudasser Habib
Nuclear Institute for Agriculture and Bi
Faisalabad, Pakistan

Bentham Science Publishers Pte. Ltd.
80 Robinson Road #02-00
Singapore 068898
Singapore
Email: subscriptions@benthamscience.net

BENTHAM
SCIENCE

CONTENTS

PREFACE

Cellular and molecular biology is a fascinating and rapidly evolving field that has revolutionized our understanding of the living world. From the intricacies of the genetic code to the complex interactions between cells, this field provides insights into the fundamental processes that make life possible.

The aim of this book is to provide a comprehensive introduction to the fundamentals of cellular and molecular biology, covering topics such as cell structure and function, DNA replication and repair, gene expression and regulation, protein synthesis and folding, and cell signaling. Through clear explanations, detailed diagrams, and engaging examples, readers will gain a deep understanding of the underlying principles and mechanisms that govern the behavior of cells and molecules.

Whether you are a student just starting in the field, or a researcher looking to review the basics, this book is an indispensable resource. With its focus on fundamental concepts and its emphasis on clarity and accessibility, it is the perfect guide to the exciting world of cellular and molecular biology.

We hope that this book will inspire readers to delve deeper into the mysteries of life and to pursue their investigations into the intricate workings of the cellular and molecular world.

Sanaullah Sajid
Institute of Microbiology
University of Agriculture
Faisalabad, Pakistan

Sajjad ur Rahman
Institute of Microbiology
University of Agriculture
Faisalabad, Pakistan

Shahid Mahmood
Institute of Microbiology
University of Agriculture
Faisalabad, Pakistan

Shayan Bashir
Premier Research, Ontario
Toronto, Canada

&

Mudasser Habib
Nuclear Institute for Agriculture and Biology
Faisalabad, Pakistan

List of Contributors

Atta Ur Rahman	Institute of Microbiology, Faculty of Veterinary and Animal Sciences, Gomal University, Dera Ismail Khan, KPK, Pakistan
Farwa Mehmood	Department of Pre-clinical and Clinical Operations, Trials 360 (CRO), Lahore, Pakistan
Hafiz Zain Ul Abideen	College of Basic and Clinical Medicine, Dali University, Dali, Yunnan, China
Haseeb Ullah Sajid	Institute of Microbiology, University of Agriculture, Faisalabad, Pakistan
Hira Asghar	Department of Science of Dental Material, Azra Naheed Dental College, Superior University, Lahore, Pakistan
Hayat Ullah	Department of Chemistry, University of Okara, Okara, Punjab, 56300, Pakistan
Itrat Fatima Toor	National Institute of Food Science and Technology, University of Agriculture, Faisalabad, Pakistan
Imran Abdullah	Institute of Nuclear Medicine and Oncology, Lahore, Pakistan
Kainat Gul	Department of Microbiology, Abbottabad University of Science and Technology, Havelian, Abbottabad, Pakistan
Munazzah Sajid	Department of Chemistry, University of Agriculture, Faisalabad, Pakistan
Muneeb Ur Rehman	Medical College of Wuhan, University of Science and Technology, Hubei, Wuhan, China
Maliha Sarfraz	Department of Zoology, Wildlife and Fisheries, University of Agriculture, Faisalabad, Sub-Campus, Toba Tek Singh, Pakistan
Maheen Shafiq	School of Engineering and Built Environment, Griffith University, Nathan, Queensland, Australia
Rai Shafqat Ali	Nuclear Institute for Agriculture and Biology, Faisalabad, Pakistan
Samiullah Sajid	Dali University, Yunnan, China
Sanaullah Sajid	Institute of Microbiology, University of Agriculture, Faisalabad, Pakistan
Shumaila Yousaf	Nuclear Institute for Agriculture and Biology, Faisalabad, Pakistan
Saba Nasir	Nuclear Institute for Agriculture and Biology, Faisalabad, Pakistan
Saima Pervaiz	Department of Immunology, University of Health Sciences, Lahore, Pakistan
Sara Masood Cheema	Department of Pathology, Azra Naheed Medical College, Superior University, Lahore, Pakistan
Shehryar Awan	Department of Pathology, Azra Naheed Dental College, Superior University, Lahore, Pakistan
Zain Ul Abedien	Institute of Microbiology, University of Agriculture, Faisalabad, Pakistan

CHAPTER 1

The Cell and its Molecular Constituents

Samiullah Sajid[1,*] and **Imran Abdullah**[2]

[1] *Dali University, Yunnan, China*

[2] *Institute of Nuclear Medicine and Oncology, Lahore, Pakistan*

Abstract: The cell is the basic unit of life and the fundamental building block of all living organisms. It is a complex and dynamic structure composed of various molecular constituents. These constituents include the plasma membrane, cytoplasm, organelles, and genetic material. The plasma membrane is a thin, flexible layer that separates the cell from its surroundings and regulates the movement of substances into and out of the cell. The cytoplasm is a gel-like substance that contains various organelles, such as mitochondria, ribosomes, and the endoplasmic reticulum, which are involved in multiple cellular functions. The genetic material, composed of DNA and RNA, contains instructions for synthesizing proteins, the building blocks of life. The cell and its molecular constituents play vital roles in maintaining the organism's integrity, responding to environmental cues, and carrying out essential physiological functions. Understanding the organization and function of these molecular constituents is crucial for advancing our knowledge of biology and developing new therapies for various diseases.

Keywords: Cell, Cytoplasm, DNA, Genetic material, Organelles, Proteins, Plasma membrane, Physiological functions, RNA.

INTRODUCTION

All living organisms are made up of small structures called cells. These structures, representing the smallest unit of life, are quite complex and diverse and contain the morphological and physiological characteristics of living organisms. The properties of a given organism depend on its cells, whose continuity occurs through their genetic material. Simple life occurs in isolated cells, which propagate by cell division. Higher organisms, on the other hand, like man himself, are made up of aggregates of cells that perform specialized functions. The cells of different organisms are very similar in structure and molecular constituents, despite fundamental organizational differences. When analyzing the molecular constituents, it is important to consider not only the individual properties of the

* **Corresponding author Samiullah Sajid:** Dali University, Yunnan, China; E-mail: hafizsamiullahsajid@gmail.com

Sanaullah Sajid, Sajjad ur Rahman, Shahid Mahmood, Shayan Bashir & Mudasser Habib (Eds.)

molecules but also the interactions between them and their location within the cell. That analysis is even more necessary when considering a multicellular organism and the events that occur inside to produce the differentiation and development of that organism. This chapter briefly reviews the cellular structure, its molecular constituents, and the interactions carried out by these compounds [1].

CELLULAR STRUCTURES

The first observations and the very name of cells for structural units, in which all functions necessary for the maintenance and preservation of life are performed, were made by Robert Hooke in 1665. Regardless of the organism's complexity, all cells have the same structure formed by the plasma membrane, which surrounds the cell contents and separates it from the extracellular environment by the cytosol and the nucleus (or nucleoid). The cytosol in the whole cell is composed of a complex aqueous solution with various dispersed particles and molecules. Cell size and shape vary greatly and have no relation to the organism's size. Some cells live in isolation, like unicellular organisms, but those of multicellular organisms generally relate to each other. The plasma membrane, which surrounds all cells, is formed by a double layer of lipids from the phospholipid class and, in variable amounts, if associated with protein molecules [2].

The plasma membrane defines the boundary between the intracellular and extracellular environment. To enter or leave a cell, a substance must cross the cell membrane, which will depend on the membrane's permeability. This lipid bilayer is permeable to certain gases, such as O_2 and CO_2, and impermeable to many substances, such as sugar, amino acids, and inorganic ions (K^+ and Cl^-); water can freely diffuse through the cell. Many proteins are bound to the plasma membrane (permeases or carriers), forming channels in the lipid bilayer and facilitating the passage of certain substances. Thus, all cells of organisms have the characteristic structural architecture of their membranes and many metabolic processes, even DNA replication, synthesis of protein, and chemical energy production.

PROKARYOTIC AND EUKARYOTIC CELLS

Despite the similarity between the cells that make living beings, organisms maintain fundamental differences at the cellular level and can be classified into two large groups: prokaryotes and eukaryotes [3]. Prokaryotic organisms are single-celled and simpler in an organization, although they occur in associated groups, forming colonies with some differentiation of functions. The prokaryotes include bacteria and archaea (bacteria that survive in unusual environments, such as saline lakes, swimming pools, thermals, and swamps). Eukaryotic organisms are more complex and include not only multicellular plants, animals, fungi, protozoa, and some single-celled organisms such as yeast and green algae (Fig. **1**).

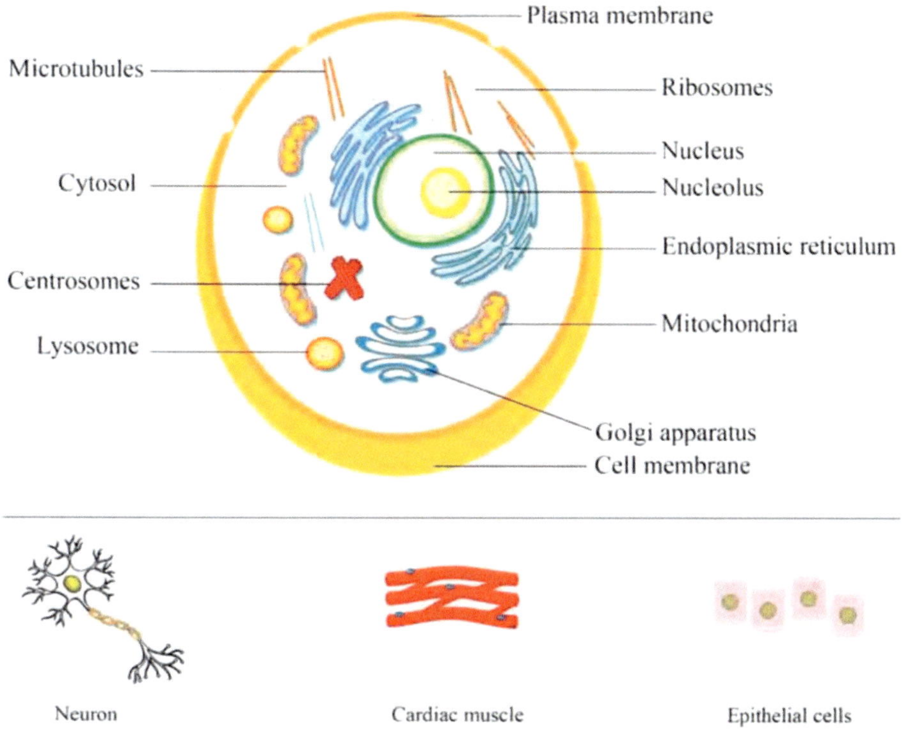

Fig. (1). Schematic representation of an animal cell (eukaryotic cell) with its main structures and cellular organelles. The main features are the presence of a well-defined nucleus containing the genome and the cell organelles, which compartmentalize certain functions.

The main difference between prokaryotes and eukaryotes is that, in eukaryotes, organelles are found, mainly the nucleus, which contains the genome. Organelles are regions bounded by internal membranes, forming specialized functions compartments. In prokaryotes, the absence of an envelope nucleus puts the genome in direct contact with the rest of the cytoplasm in a space within the cell called the nucleoid, staying next to ribosomes, other particles, and a wide variety of dissolved molecules. Prokaryotic cells usually have a cell wall, whose function is to provide greater rigidity and mechanical protection. This membrane is also present in eukaryotic plant cells. The chemical composition of the cell wall of prokaryotes is quite complex, containing molecules of polysaccharides, lipids, and proteins.

In eukaryotic cells, the Nuclear DNA is split into two or more chromosomes. Each of these chromosomes comprises a linear DNA molecule that, except during cell division, is confined within the nucleus. These DNA molecules are associated with proteins, called histones, forming nucleosomes and components of chromatin. The number and size of individuals' chromosomes vary greatly among different eukaryotic organisms. Fungi, for example, has 12 to 18 chromosomes; human cells contain two sets of 23 chromosomes, each about thirty times the amount of DNA in a cell of the *Escherichia coli* bacterium.

Organelles are common to all eukaryotic cells, the nucleus being perhaps the most important of these organelles. The nucleus is surrounded by a double nuclear membrane, which has a characteristic, in addition to the chromatin, a region rich in RNA, called the nucleolus. In the nucleolus, RNAs are synthesized from a DNA template and exported to the cytoplasm across the nuclear membrane. Mitochondria, another important organelle, has enzymes specialized in oxidative processes that produce energy for the cell. In addition, DNA and ribosomes are found inside the mitochondria. The endoplasmic reticulum comprises membranes distributed throughout the cytoplasm and bound to the cell and nuclear membranes. The granular ER has ribosomes attached to its membranes, constituting the molecular machinery for protein synthesis. The Golgi complex is formed by membranes and vesicles, involved in the modification and secretion of proteins from cells. There are specific organelles for plant and animal cells. Animal cells, for example, contain lysosomes with digestion function, and plant cells have chloroplasts, where photosynthesis occurs. Another feature common to most cells of plants and some microorganisms is the presence of a vacuole, whose function is to store nutrients and metabolites. Each organelle has its collection of enzymes that catalyze specific reactions, developing a unique role in cell growth and metabolism. Cell organelles are linked to a network of microtubules of filamentous proteins called the cytoskeleton. This structure is responsible for maintaining cell shape and assisting in cellular movements.

Molecular Constituents

Molecular constituents are responsible for the biochemical interactions between thousands of molecules that enable cellular life. These chemical reactions occur in an aqueous medium, so water, with few exceptions (bone cell), is the component found in greater amounts in the cell, indispensable for metabolic activity. Due to its polar nature, water serves as a natural solvent for ions, minerals, and other substances and as a dispersion medium for the colloidal structure of the cytoplasm. The presence of ions, such as Cl^-, Na^+, and K^+, is important for maintaining the cell's osmotic pressure and acid-base balance. Some inorganic ions, such as magnesium, are required as enzymatic cofactors, while others, such as inorganic phosphate, form adenosine triphosphate (ATP), the primary chemical energy source of vital processes, and calcium ions play a regulatory role.

In addition to water and the chemical elements mentioned above, the cell comprises small molecules and macromolecules. Small molecules such as amino acids, nucleosides, lipids, and sugars constitute the substrates and the products of metabolic pathways, providing energy for the cell and may also be the forming units of macromolecules. In this biological polymer structure, these molecules are called monomers. Cells comprise three polymers: nucleic acids formed by nucleotides (monomers); proteins, made up of amino acids; and carbohydrates or polysaccharides, whose monomers are sugars or monosaccharides. Cells have many lipids, not polymers, preferably small molecules, unlike the others. The simplest and most abundant lipid is the fatty acid, which participates in the composition of other more complex lipid molecules, such as triacylglycerols. Larger structures, such as the bilayer of biological membranes and the cytoskeleton fibers, are formed by the association of these macromolecules.

Amino Acids and Proteins

Proteins result from the expression of the information contained in the gene. Therefore, the gene will determine the amino acid sequence of a specific protein. Like this, every protein has a definite order of residues. Amino acids establish their active three-dimensional structure and are called native conformation. The three-dimensional structure of the molecule, established according to the amino acid sequence itself, can be observed in denaturation experiments of a given protein. The amino acid activity depends on the changes in conditions of the environment (changes in pH, temperature, addition of solvents) where the protein is found, with a consequent loss of their biological function, can sometimes be recovered.

Restoration of the native conformation (protein renaturation) can occur when the conditions of the environment in which the molecule is found are reestablished, allowing the amino acids to return to interact. The folding of a globular protein is an energetically favourable process under physiological conditions that enable interactions between chemical groups. Proteins are classified into two main classes: fibrous and globular. Fibrous proteins play a structural role in cells and animal tissues. This class includes collagen, a component of bones and connective tissue, and keratin in nails and hair. Globular proteins have a spiral and compacted structure with a globular shape; abundant and essential, they can be found in any organism. One example is enzymes, efficient biological catalysts that speed up chemical reactions. Except for some RNAs (ribozymes), which have catalytic activity, all enzymes are proteins. All globular proteins have a unique, coiled structure in a specific way and according to the particular function to be executed. Since the structure of a protein determines its function, it is important to know the structural characteristics of this molecule.

Proteins are long chains of amino acids and constitute more than half of the dry weight of a cell. They are also polymers that perform numerous biological functions in addition to determining the shape and structure of the cell. Proteins are also known as molecules that do cellular work. They catalyze an unexpected number of chemical reactions, control the permeability of membranes, regulate the concentration of metabolites, recognize and non-covalently bind other biomolecules, provide movement, and control gene function. Proteins consisting of only 20 amino acids joined by peptide bonds perform this diverse number of functions.

Like the native conformation, which allows the protein to perform its functions, is a consequence of the individual properties of the amino acids present in the protein molecule, it is important to review these properties. Amino acids are named after organic acids, with a carbon atom (C) attached to four chemical groups. An amino group (NH_2), a carboxylic group (COOH), a hydrogen atom (H), and a variable group are called side chains or radicals (R). Except for glycine, which also has a hydrogen atom in the radical, all other amino acids have four different groups attached to C, giving rise to an asymmetric carbon. The presence of this asymmetric carbon generates two molecules (stereoisomers) called D (dextro) and L (levo) isomers. At physiological pH, the amino and carboxylic groups of amino acids are ionized (NH3+ and COO-), causing the amino acid to have positive and negative charges in the same molecule (dipolar molecule).

The simultaneous presence of these groups determines the acid-base behavior of amino acids (amphoteric molecules). The acid-base character and the amino acid's electrical charge are determined by the pH of the medium where it is found. The side chains of amino acids vary in size, shape, electrical charge, hydrophobicity, and reactivity. Amino acids with polar side chains are hydrophilic and tend to be located on the protein's surface due to their interactions with water. Furthermore, the more polar amino acids are present in the protein, the more soluble it will be in aqueous solutions. However, because they are hydrophobic, amino acids with non-polar R groups tend to be inside the proteins and cause their insolubility in water. Arginine and lysine (basic polar) are positively charged in the polar group, and glutamate and aspartate (acidic polar) are negatively charged. Four amino acids are primarily responsible for protein loads; histidine, which also has a positive charge, helps maintain the pH (physiological buffer) due to its ability to capture or release protons through the imidazole group in the radical. Belonging to the class of neutral polar amino acids, cysteine can react with other cysteine residues using the thiol group (SH), present in the para radical disulfide form (S–S) bridges in an oxidation reaction, and this link plays an important role in the conformation of proteins. Amino acids with hydrophobic side chains are almost insoluble in water due to the presence of hydrocarbons in these groups. Phenylalanine, tryptophan, and tyrosine are aromatic groups responsible for the ultraviolet light absorption of proteins at 280 nm. Proline is a unique amino acid since its side chain is covalently bonded to the nitrogen of the amino group, forming a rigid ring. The presence of proline in a protein chain can restrict the way the molecule will fold.

During the synthesis of the protein molecule, the amino acids are united by a covalent bond between an amino acid's carboxylic group and another amino acid's amino group, connected by a dehydration reaction with the loss of a water molecule. The molecule formed generates a peptide and maintains its amphoteric character since there will always be a free carboxylic group at one end (C-terminus) and a free amino group at the other (N-terminal) end. The combination of just two amino acids forms a dipeptide; the union of a few amino acids gives rise to oligopeptides, while a polypeptide comprises many amino acids (sometimes greater than 1,000). The sequence of a protein chain is written with the N-terminus on the left and the C-terminus on the right. The size of a protein is, overall, expressed by its mass in Daltons (Da). Enzymes form an important class of protein that catalyzes all chemical reactions. When involved in redox reactions, some are only active when covalently linked to a coenzyme (prosthetic group), such as nicotinamide adenine dinucleotide (NAD+), whose structure is formed by a nicotinamide ring, an adenine ring, and two linked phosphate sugar groups.

Three-dimensional Protein Structure

The formation of a polypeptide chain, considering the correct polymerization of amino acids, is carried out in the translation process and determined by the information in messenger RNA (mRNA). The protein chain synthesized assumes a precise spatial organization and is necessary for the protein to perform its function (native conformation). The three-dimensional structure of a protein is the combination of several factors, mainly interactions between chemical groups present in this protein and stereochemical limitations imposed by the peptide bond itself due to the water resonance character. This resonance prevents the rotation of the carbon attached to the nitrogen, leaving all the atoms involved in the peptide bond in the same position. Proteins are analyzed considering the four levels of structural organization. The primary structure is the first step in specifying the structural analysis of a protein and refers to the sequence of amino acids, that is, the order in which the amino acids are linked to form a peptide chain. Each protein has its specific primary structure, which, in turn, determines the three-dimensional structure. The biological importance of the amino acid sequence is well exemplified in the hereditary human disease called sickle cells. In this disease, profound biological changes occur, caused by replacing a single amino acid in the hemoglobin molecule.

Secondary structure refers to the various spatial arrangements of amino acids close in the central peptide chain, which causes folding; such folds are called secondary structures. These arrangements may feature an organization that repeats at regular intervals. The most common secondary structure organizations are the a-helix in which the peptide chain winds around an axis, are stabilized by hydrogen bonds formed between the amino group of the peptide bond of an amino acid and the carboxyl group located four residues ahead in the same polypeptide chain.

Tertiary structure refers to how the polypeptide chain is coiled, including the three-dimensional arrangement representation of all atoms in the molecule, including those in the side chain and the prosthetic group. This level of structure is established when different secondary structures are arranged among themselves. The structure's stability is maintained by hydrogen bridges between R groups, hydrophobic interactions, ionic bonds between positively and negatively charged groups, and disulfide (S–S) covalent bonds. In globular proteins, the side chains of the more hydrophobic amino acids aggregate inside the molecule, and the groups of hydrophilic proteins protrude from the surface of the protein. The final three-dimensional structure can comprise a specific combination of secondary structures and helices entwined, forming compactly folded globular units called domains. The tertiary structure of larger proteins is subdivided into domains; these domains have about 100 to 150 amino acids and are linked by the peptide chain. Proteins with more than one polypeptide, forming subunits (multimeric proteins), have a quaternary structure, and this structure refers to the arrangement of the protein subunits that make up the molecule. The number of subunits can vary, and the union between them occurs in a non-covalent way through electrostatic interactions with hydrogen and hydrophobic bonds. Some proteins, called allosteric, exhibit a cooperative effect between the subunits so that a change in one of these subunits can result in a change in another subunit. A good example is the tetrameric hemoglobin molecule composed of four polypeptide chains forming the subunits. Each subunit binds to an oxygen molecule cooperatively; that is, after a molecule of oxygen is bound to one subunit, the binding of the other molecules is facilitated.

Carbohydrates

Carbohydrates, or monosaccharides, are simple sugars and represent one of the large classes of biological molecules with various cellular functions. Polysaccharides are polymers with long chains of monosaccharide units and constitute the primary cellular energy source. They are also important structural features of the constituents of the cell wall, acting as specific recognition signals and playing an informational role. Carbohydrates are formed by carbon covalent bonds, in a 1:1 ratio, and water (CH_2O). They are sorted according to the number of carbon atoms present in the molecule: trioses (3), pentoses (5), or hexoses (6). All monosaccharides can contain hydroxyl groups and an aldehyde or ketone group. Those two groups can react with a hydroxyl group in the same molecule through a reaction, converting a linear structure to a ring-shaped one (Fig. **2**).

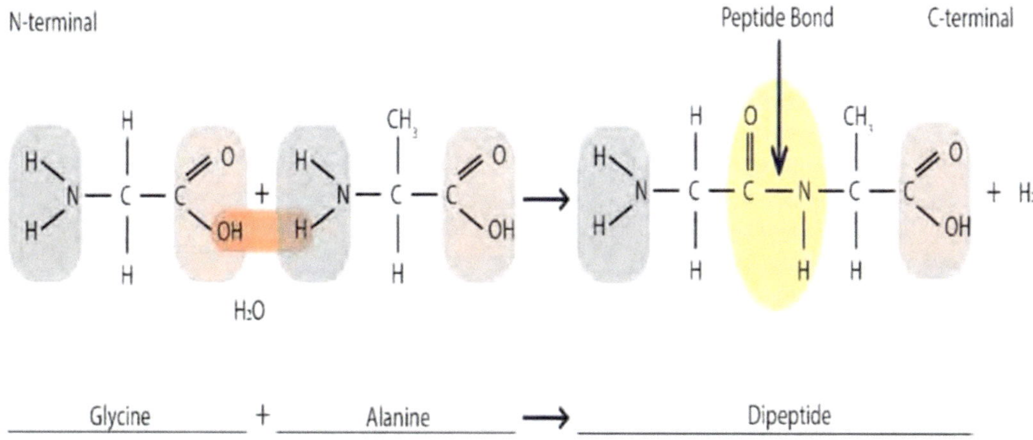

Fig. (2). Schematic drawing of the formation of a peptide through the bond between the carboxylic group of one amino acid and the amino group of another (peptide bond).

Glucose is a monosaccharide containing six carbon atoms and an aldehyde group. An example is D-glucose, the primary source of energy for most cells. The structure of D-glucose can be presented as a linear chain with two different structures. When the aldehyde group on carbon 1 reacts with the hydroxyl group on carbon 5, the resulting ring has six elements, generating a D-glucopyranose. If the hydroxyl group occurs with carbon 4, the structure is a D-glucofuranose, whose presence in nature is much rarer. All monosaccharides except dihydroxyacetone contain one or more asymmetric carbons, generating optically active stereoisomers (D and L). The cyclization of the linear structure generates new isomers, called anomers, because they are linked to anomeric carbon. Oligosaccharides are molecules formed mainly by linking a few monomeric units. An example is sucrose, a disaccharide formed by the union of a glucose molecule and a fructose that produces the common sugar used in food after being processed.

The most important polysaccharides in living organisms are starch and glycogen, as they represent reserved substances and energy storage in plants and animal cells. Glycogen is a polysaccharide formed by linking several glucose molecules. Cellulose is also an important polysaccharide and is the main structural element of the plant cell wall. Disaccharides, like polysaccharides, are made up of monosaccharides joined covalently by glycosidic bonds. These bonds are formed when a hydroxyl group on a carbohydrate's anomeric carbon reacts with another carbohydrate's hydroxyl group. Those free hydroxyl groups can also bind with other amino groups, sulfate, and phosphate of different molecules, forming more complex molecules, such as glycosaminoglycans, the main components of the extracellular matrix.

Lipids

Lipids form a distinct group of compounds with multiple cellular functions frequently occurring in nature. They are usually molecules with a strong tendency to associate with each other through non-covalent forces, forming lipid aggregates. Lipids are generally characterized by their fatty acid structure. One fatty acid molecule has two distinct regions: a polar, hydrophilic region connected to a non-polar hydrophobic region consisting of a hydrocarbon chain. This structure characterizes lipids as poorly water-soluble compounds but soluble in organic solvents. This molecular feature promotes an amphipathic-like association of lipid molecules with non-covalent interactions in an aqueous medium. These interactions have considerable consequences at the cellular level; the most important of these is the tendency of lipids to form micelles and bilayers, which make up biological membranes. The exact structure formed when the lipid is in contact with water depends on the specific molecular structure of the hydrophilic and hydrophobic regions of the molecule.

Fatty Acids

The simplest lipids are fatty acids, also constituents of more complex lipids. Its basic structure represents most lipid molecules in large amounts in human cells. The structure of the fatty acid is formed by a long hydrocarbon, hydrophobic, and little chemically reactive chain. In general, the fatty acids found in living organisms contain even more carbon atoms, and their hydrocarbon chain is unbranched. Fatty acids are classified as saturated, unsaturated, or polyunsaturated, depending on the bonds between the carbon atoms. In saturated fatty acids, the chain contains only single bonds; if there are double bonds, the fatty acids are unsaturated, while fatty acids with more than one double bond are called polyunsaturated. Two polyunsaturated fatty acids, classified as essential acid, linoleic acid, with 18 carbons and two double bonds, and linolenic acid, also with 18 carbons but with three double bonds. An extremely hydrophilic, ionizable carboxylic group forms the fatty acid molecule in solution (COO-), including triacylglycerols (triglycerides) fatty acid. Triacylglycerols are the storage form of lipids in the cytoplasm of many cells because they serve as an excellent power source due to the presence of the carbon chain. Thus, triacylglycerols are more efficient energy storage than carbohydrates and are widely used by various organizations, including higher animals.

Phosphoglycerates

Phosphoglycerates or phospholipids are small lipid molecules composed of long chains of fatty acid and glycerol linked to a highly polar group. They differ from triacylglycerols in having only two fatty acid molecules joined to a single glycerol molecule, in which a third hydroxyl is esterified to a phosphoric acid (phosphatidic acid). This phosphate can be linked to a hydrophilic molecule (choline, ethanolamine, inositol, or serine), depending on the type of phosphoryl. The amphipathic nature of Phosphoglycerates is responsible for the molecular associations that form the cell membrane and confer many of its properties [4].

Steroids

Steroids are a large group of molecules that aggregate several functions, including many hormones, like the sex hormones of higher animals. Cholesterol is the most important steroid, part of cell membranes, mainly from animals. The steroids derive from a general structure containing three rings of 6 carbons (A, B, and C) and a ring with five atoms. The cholesterol molecule is not very amphipathic, as a hydroxyl group is located at the end of the molecule. The fused cyclohexane rings of this molecule form a very rigid structure, and their presence in the membrane tends to break the regularity of the structure, giving it greater rigidity. This compact structure is also responsible for harmful health effects, such as atherosclerosis and cardiovascular disease, caused by cholesterol storage in the blood vessels.

Nucleic Acids

Nucleic acids are macromolecules of great biological importance in all living organisms. From nucleic acids, cells receive information about proteins to synthesize, the sequence of amino acids in their structure, and the function of these molecules. They are the molecules that store and transmit genetic information in the cell. All this information in genes located on the chromosomes of cells is deciphered through the genetic code, whose translation results in protein synthesis. There are two types of nucleic acids: deoxyribonucleic acid (DNA) and ribonucleic acid (RNA); both are linear polymers of nucleotides joined by phosphodiester bonds.

In most cases, the number of monomers in a nucleic acid molecule is much larger than the number of amino acids in a protein molecule. RNAs vary in size and may have ten to thousands of nucleotides. DNA and RNA consist of only four different types of nucleotides. Each nucleotide comprises a group phosphate, a sugar (pentose), and a nitrogenous base (purine or pyrimidine) linked by covalent bonds. The differences between the two nucleic acids reside in the type of sugar and composition of the base of the molecule. In RNA, pentose is always ribose, and in DNA, it is deoxyribose. The fundamentals are adenine (A), guanine (G), and cytosine (C), found in both DNA and RNA; the base thymine (T) is present only in DNA, and uracil (U) in RNA. These letters (A, C, G, T, and U) indicate a sequence of nucleotides in a nucleic acid. In addition to these five bases, there are unusual bases with slightly different structures, which occur mainly in RNA, such as hypoxanthine and inosine, to link sugar and the phosphate group. A phosphate group replaces the hydroxyl group of C5 of the sugar, and without the phosphate group, a base and sugar combination constitute a nucleoside. When nucleosides polymerize to form a nucleic acid, the hydroxyl group attached to the C3 of the sugar of one nucleotide forms an ester bond with the phosphate of another (phosphodiester bond), thus eliminating a water molecule and forming a backbone of pentose phosphate repeating units with the bases attached to the side group. On a strand of DNA or RNA, at one end, there is a cluster bonded phosphate to the sugar's 5' carbon; at the other end, there is a hydroxyl group attached to the 3' carbon. The polynucleotide chain has individuality, determined by the sequence of its bases, known as the primary structure, that contain genetic information to a particular sequence of DNA (protein or RNA).

Deoxyribonucleic Acid

Deoxyribonucleic acid (DNA) [5] is found in living organisms as molecules of high molecular weight. For example, *E. coli* has a single circular DNA molecule of 4.2×10^6 bp and a total length of 1.4 mm. The amount of DNA in higher-level organisms can be hundreds of times larger (700 times in the case of a man); the DNA of a single fully extended diploid human body cell can be up to 1.7 m long.

All the genetic information of an organism is accumulated in the linear sequence of the four bases. The primary structure of all proteins (20 amino acids) must be encoded by a four-letter alphabet (A, T, G, and C). Between 1949 and 1953, Chargaff, studying the base composition of DNA, demonstrated that although the composition varied from one species to another, in all cases, the amount of adenine was equal to that of thymine (AT), and that of cytosine was equal to that of guanine (CG). Thus, the total number of purines was equal to that of pyrimidines (A+G, C+T). On the other hand, the AT/GC ratio varied considerably between the species. In 1953, based on X-ray diffraction data, Watson and Crick proposed a model for the structure of DNA. This model explained the regularities of base composition, mainly its duplication in the cell. DNA is composed of two associated polynucleotide chains, which coil to form a double helix around a central axis that rotates to the right most of the time. The bases are inside the helix, perpendicular to the helical axis, interacting through hydrogen bonds that unite the two chains. The only possible pairings are A/T and C/G, and it is important to note that two hydrogen bonds are present between A and T, and between C and G, three bonds; as a result, the G/C pair is more stable than the A/T. In addition to these hydrogen bonds, hydrophobic interactions stabilize the double helix. The orientation of the two types of DNA is antiparallel; that is, the 5' → 3' direction of each one is opposite to the other.

Ribonucleic Acid

Ribonucleic acid (RNA) [6] is a nucleic acid consisting, in general, of a single strand with a great diversity of conformations. The sequence of bases (primary structure) is similar to that of DNA, except for replacing deoxyribose with ribose and thymine with uracil. There are three main classes of ribonucleic acid: messenger RNA (mRNA), which contains the genetic information for the sequence of amino acids; transfer RNA (tRNA), which identifies and transports amino acid molecules to the ribosome; and ribosomal RNA (rRNA), which accounts for 50% of the mass of ribosomes. All types of RNAs are involved in protein synthesis. Although the RNA molecule comprises only one polynucleotide chain, its structure is complex. The molecule adapts in a helical structure similar to DNA in paired regions. Differences in the size and conformations of the various types of RNA allow them to perform specific functions. The tRNAs have a conformational structure that allows pairing the anticodon nucleotides in the molecule with the mRNA codon nucleotides.

GLOSSARY

<u>Cell:</u> The basic unit of life, consisting of a plasma membrane, cytoplasm, and a nucleus or nucleoid.

<u>Plasma Membrane:</u> The selectively permeable outer boundary of a cell, composed of a phospholipid bilayer with embedded proteins and carbohydrates.

<u>Cytoplasm:</u> The gel-like fluid that fills the cell, containing organelles, cytoskeleton, and other macromolecules.

Nucleus: The organelle in eukaryotic cells that contain genetic material in chromosomes.

Nucleolus: A sub-organelle within the nucleus that is responsible for the synthesis of ribosomal RNA.

Chromosome: A long, coiled-up structure of DNA and protein that carries genetic information.

Ribosome: An organelle synthesizes proteins by translating mRNA into amino acid sequences.

Endoplasmic reticulum (ER): An organelle composed of a network of membranous tubules and sacs that plays a role in protein synthesis, lipid metabolism, and calcium storage.

REVIEW QUESTIONS

1. What is the plasma membrane, and what is its function?

2. What is the role of the endoplasmic reticulum in a cell?

3. What is an enzyme, and how does it function in a cell?

REFERENCES

[1] Alberts B, Johnson A, Walter P, Lewis J, Raff M, Roberts K. Molecular cell biology. New York: Garland Science 2008.

[2] Campbell MK, Farrell SO, Tasks A. Biochemistry. 5., They are Paul: Thomson 2007.

[3] Devlin TM. Biochemistry manual with clinical correlations São Paulo: Edgard Blucher; 2007 Lewin B Genes IX 9. Porto Alegre: Artmed 2009.

[4] Lodish H, Baltimore D, Berk A, Zipursky L, Matsudaira PT. Cellular and molecular biology 5. Porto Alegre: Artmed 2005.

[5] Persing DH, Ed. Molecular microbiology: Diagnostic principles and practice. Washington: ASM Press 2004.

[6] Alberts B. Molecular Biology of the Cell. New York: Imprint W.W. Norton & Company 2015. [http://dx.doi.org/10.1201/9781315735368]

CHAPTER 2

Nucleic Acid Structure

Atta Ur Rahman[1,*]

[1] *Institute of Microbiology, Faculty of Veterinary and Animal Sciences, Gomal University, Dera Ismail Khan, KPK, Pakistan*

Abstract: Nucleic acids are essential biomolecules that store and transmit genetic information in all living organisms. The structure of nucleic acids, specifically DNA (deoxyribonucleic acid) and RNA (ribonucleic acid), is crucial for their function. The backbone of nucleic acids is composed of alternating sugar and phosphate groups, while the nucleotide bases (adenine, guanine, cytosine, thymine and uracil) protrude from the sugar-phosphate backbone. The bases form specific hydrogen bonds with complementary bases on the opposite strand, resulting in the double-helical structure of DNA. RNA is usually single-stranded but can form secondary structures through base pairing. The three-dimensional structure of nucleic acids is also important for their function, as it determines the interactions between nucleic acids and other molecules, such as proteins. Understanding nucleic acid structure is fundamental for many areas of biology, including genetics, molecular biology and biotechnology.

Keywords: Cellular process, Gene expression, Genetic information, Nucleotide sequence, Polynucleotide synthesis, Ribonucleic acid, Ribosomal RNA, Transfer RNA.

INTRODUCTION

DNA and RNA are crucial nucleic acids that store and transfer genetic information in living organisms. DNA, a double-stranded molecule, acts as the blueprint for directing cellular functions. Cells transcribe the instructions encoded in DNA into the nucleotide sequence of single-stranded RNA molecules. RNA molecules have various tasks, including the creation of polypeptides, regulation of gene expression, and defense against invading nucleic acids, such as those encountered in viral infections. The study of nucleic acid structure and function, which began over 60 years ago, has provided an unprecedented understanding of biological processes and has transformed several professions, including illness detection, treatment, and forensic inquiry [1].

* **Corresponding author Atta Ur Rahman:** Institute of Microbiology, Faculty of Veterinary and Animal Sciences, Gomal University, Dera Ismail Khan, KPK, Pakistan; E-mail: arehman@gu.edu.pk

Sanaullah Sajid, Sajjad ur Rahman, Shahid Mahmood, Shayan Bashir & Mudasser Habib (Eds.)

Throughout history, humans have observed patterns of heredity without grasping the underlying mechanisms responsible for conveying physical features and developmental processes from parents to offspring. However, these discoveries have been leveraged by diverse societies to strengthen their economic conditions through selective breeding and crop production. The scientific area of genetics began in the nineteenth century, dedicated to solving the secrets of heredity. At the start of the twentieth century, scientists largely recognized that physical attributes are inherited through discrete units known as genes, which sit on chromosomes within the nucleus of cells. After decades of tedious investigation, the chemical makeup of chromosomes was revealed, leading to the identification of DNA as the bearer of genetic information. The crucial discovery of the DNA structure by James Watson and Francis Crick in 1953 marked a turning point and gave rise to the science of molecular biology.

Molecular biology is a specialist field that focuses on researching the structure of genes and the complicated mechanisms involved in DNA manipulation. Researchers in life sciences, armed with cutting-edge technology developed by molecular biologists and biochemists, have been actively examining the manner in which living creatures organize and process genetic information. This huge corpus of efforts has uncovered numerous essential principles:

1. DNA serves as the central director of cellular operations and is inherited by offspring. It consists of two polynucleotide strands that interlock to form a double helix. The information stored in DNA is delivered by the unique sequence of purine and pyrimidine nucleotides. A gene is a particular sequence of DNA that comprises the required information, including both regulatory sequences that govern gene product synthesis and the instructions for encoding a gene product, such as a polypeptide or various types of protein molecules. DNA replication, known as the synthesis of new DNA strands, includes the precise matching of purine and pyrimidine bases between the original DNA strand and the newly produced strand. The functioning of DNA, both physiologically and genetically, relies on the synthesis of accurate copies of the original and pristine DNA strands. To maintain the fidelity of genetic information, various DNA repair mechanisms are employed by most organisms.
2. The process by which genetic information is decoded and utilized to orchestrate cellular processes commences with the synthesis of another class of nucleic acid called ribonucleic acid (RNA). RNA synthesis, known as transcription, involves the complementary pairing of ribonucleotides' bases with the bases of a DNA molecule. Each newly synthesized RNA molecule is termed a transcript, and the entirety of RNA molecules transcribed from an organism's genome is referred to as the transcriptome.

3. Several types of RNA directly participate in enzyme synthesis, while other proteins are indispensable for the regulated production of all other biomolecules essential for an organism's function. Messenger RNA (mRNA) contains base sequences that specify the primary structure of a particular polypeptide. Ribosomal RNA (rRNA) molecules serve as crucial components of ribosomes, cellular structures responsible for protein synthesis. Each transfer RNA (tRNA) molecule is covalently attached to a specific amino acid and transports it to the ribosome for incorporation into a growing polypeptide chain. The process of protein synthesis, termed translation, transpires within ribosomes, which are molecular machines composed of ribonucleoproteins that interpret the base sequences of mRNA molecules and translate them into amino acid sequences of polypeptides. The complete set of proteins synthesized by a cell is known as the proteome.
4. Gene expression encompasses the mechanisms through which cells temporally regulate gene synthesis in response to environmental or developmental signals. A vast array of proteins known as transcription factors and RNA molecules called non-coding RNA (ncRNA) molecules modulate gene expression by binding to specific DNA sequences. The metabolome refers to the comprehensive collection of low-molecular-weight metabolite molecules generated by a cell as a consequence of its gene expression pattern.
5. The flow of genetic information can be summarized by the central dogma, which illustrates the sequence of molecular events. DNA is transcribed into RNA, and the RNA molecules are translated into proteins. This fundamental principle initially proposed that genetic information flows in one direction only: from DNA to RNA to proteins. However, a significant exception to the central dogma was discovered in recent years. Certain viruses possess RNA genomes and possess an enzyme called reverse transcriptase. Upon infecting a host cell, reverse transcriptase copies the viral RNA to generate a DNA copy, which is then inserted into the host chromosome. An example of such a virus is HIV, as described in the perspective on "HIV Infection".

The present chapter is dedicated to exploring the structure of nucleic acids. It commences with a detailed description of DNA structure and the scientific investigations that led to its discovery. Subsequently, it provides an overview of the current understanding regarding the structure of the genome and chromosomes, as well as the structure and functions of different forms of RNA. The chapter concludes by elucidating the macromolecular complex formed by nucleic acids and proteins, which are cellular parasites.

The study of nucleic acids has yielded profound insights into the intricate mechanisms of life. It has unraveled the secrets of genetic information storage and transmission, fueling practical applications across diverse fields. The understanding of nucleic acids has opened up new avenues in disease diagnosis, treatment, and forensic investigations, revolutionizing our approach to biology and illuminating the wonders of the genetic code.

DEOXYRIBONUCLEIC ACID (DNA)

Deoxyribonucleic acid (DNA) [2] is a polymer made up of deoxyribonucleotide units. It is a very long macromolecule; for term comparison, while a typical eukaryotic cell is about 50 meters long, the smallest human chromosome, if extended, is approximately 14,000 meters long; that is, a single human chromosome is 280 times longer than the diameter of the cell that contains it. Deoxyribonucleotides comprise sugar, a nitrogenous base, and phosphate groups (PO4-). The sugar is the pentose deoxyribose (2'-deoxy-D-ribose), linked at carbon 1' to a heterocyclic nitrogenous base through 5' carbon to phosphate groups. Nitrogenous bases can be of two types: the purine bases, which are adenine (A) and guanine (G), and the pyrimidine bases, which are cytosine (C) and thymine (T). The bond between the nitrogenous base and pentose is called a glycosidic bond. The molecule composed of the sugar-bound nitrogenous base, without phosphate groups, is called a nucleoside. When a phosphate group is attached to the 5' carbon of the pentose, the molecule is called a nucleotide, which can be mono-, di- or tri-phosphate, depending on the number of phosphates linked to it. The first phosphate group attached to the nucleotide is called (alpha), the second (beta), and the third (gamma). The molecule of DNA will always contain mono-phosphate nucleotides, and the precursors for their synthesis will always be triphosphates. The deoxyribonucleotides are denominated according to the nitrogenous base and the number of phosphate groups attached to the pentose.

In the DNA molecule, the deoxyribonucleotides form chains, joined by covalent connections, forming phosphodiester bridges between the phosphate grouping (5'-PO4, or simply 5'-P) and the hydroxyl group (3'-OH) of the 3' carbon of the adjacent nucleotide. This gives polynucleotide chains an important property, which is directionality. The "first" nucleotide of the chain will have a 5' end with a phosphate grouping, and the "last" nucleotide at the 3' end will have a hydroxyl group. The following nucleotide to be added to the chain will mandatorily be added at the 3'-OH end of the polynucleotide chain. Polynucleotide chains are, by convention, represented in the 5'→3' orientation, and only the letters indicative of the nitrogenous bases are represented, as in the example:

5'-AACGTTGCTATCGT-3'

Detailed studies on the crystallographic structure of DNA have shown that the angles of covalent bonds and the interactions between the electron clouds confer certain rigidity to the spatial conformation of the chains of the DNA, even when only one of the chains (strands) is formed. The highest degree of rotation occurs in the bonds between the oxygen and the phosphorus (phosphodiester bond) and between the pentose and the nitrogenous base (glycosidic bond). However, there are preferential locations even at these positions, demonstrating that the DNA molecules assume non-random spatial conformations. These chemical characteristics of the DNA chains contribute to the formation of the double helix (Fig. **1**).

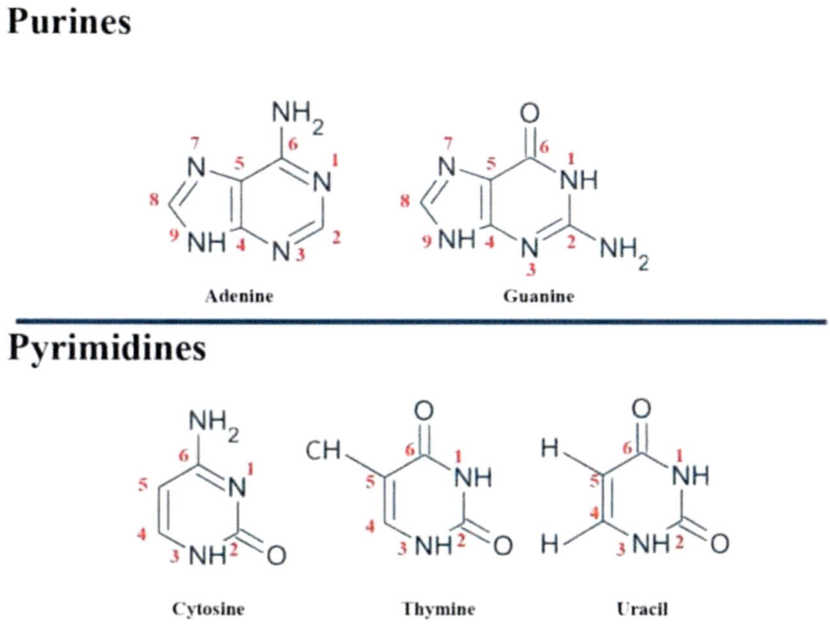

Fig. (1). Structure of Purines and Pyrimidines.

Double Helix Model

The elucidation of the secondary structure of the DNA molecule was fundamental in its definition as a molecule that stores genetic information. The principal property of this function is the capacity for "self-replication." In addition to many studies on the chemistry of DNA, the use of X-ray diffraction methods was necessary for the discovery of the structure of DNA. The diffraction permits the spatial location (three-dimensional, in the x, y, and z axes) of the atoms in a crystallized molecule. It thus defines the spatial conformation of the molecule. In 1953, James Watson and Francis Crick proposed a model of the three-dimensional structure of DNA based on X-ray diffraction studies conducted by Rosalind Franklin and Maurice Wilkins and in other chemical analyses of the molecule.

The model proposed by Watson and Crick showed that DNA is a double helix and that its two strands wrap around the axis of the helix. The deoxyribose bound to each other by phosphate groups is external to the nitrogenous bases as if they were like the handrail of a circular staircase, exposed to the aqueous medium. The phosphodiester bonds in the two strands are in opposite directions - one strand in the 5'→ 3' direction and the other in the 3'→ 5' direction and are therefore said to be antiparallel. The pairing of the bases is crucial for maintaining the double helix. The nitrogenous bases' aromatic rings are hydrophobic and oriented double helix, almost perpendicular to their axis as if they were the "rungs" of a ladder. Each nitrogenous base in one of the chains is paired with the complementary base in the other chain of DNA (Fig. **2**). This pairing up occurs by forming hydrogen bonds between the two strands, maintaining the molecule's structure.

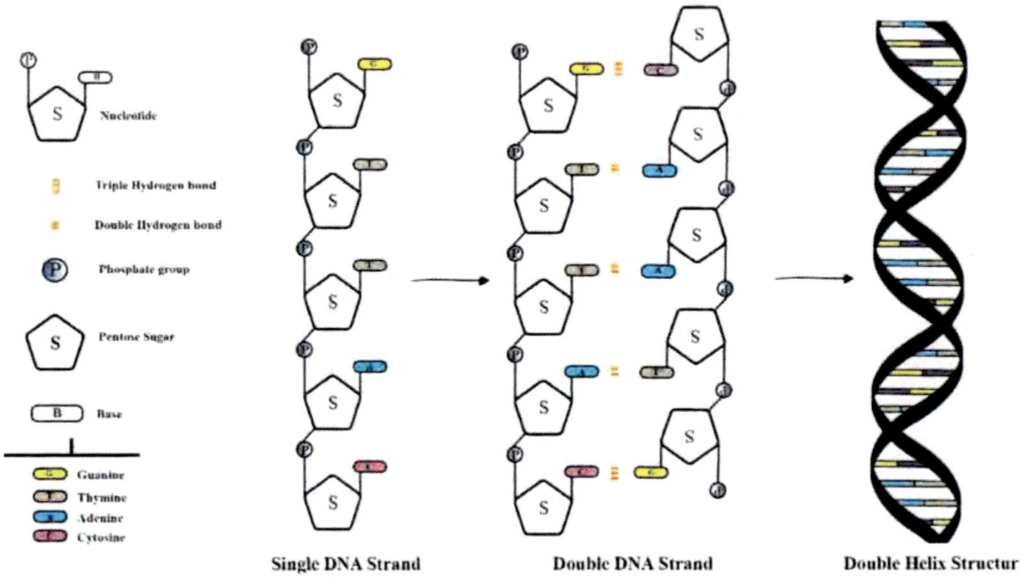

Fig. (2). Double helix structure of DNA.

A hydrogen atom usually forms a covalent bond with just one other atom. However, if this bond occurs with an atom that is an electronegative donor, hydrogen can form additional weak associations, such as hydrogen bonds, with another acceptor atom. The presence of keto (CO) and amine (C-NH2) groups allows the formation of hydrogen bonds between the bases. From this, T and U, which contain a keto group, can pair with A, which includes an amino group, through a hydrogen bond. C and G can form two hydrogen bonds with a keto and an amino group.

Furthermore, an additional hydrogen bond can be formed between the nitrogen of the aromatic rings in all pairs. Thus, between T or U and A, two hydrogen bonds are formed, and between C and G, three hydrogen bonds are formed. This confers to the molecule of DNA complementarity; that is to say, always whenever there is an A in one strand, there will be a paired T, and when there is a C in the other strand, there will be a G. This is a fundamental property of DNA and the basis for replication, transcription, and other cellular processes.

Nitrogenous bases have two types; pyrimidine is smaller than purines. Pairing always occurs between a purine base and a pyrimidine base; thus, AT and CG pairs are approximately the exact sizes and similar dimensions. Therefore, the two pairs occupy the same three-dimensional space, conferring uniformity throughout the DNA molecule. For this reason, there is no restriction on the sequence of nucleotides in the DNA molecule. These pairing characteristics explain that, in any double-stranded DNA sequence, the relative molar ratio between A/T equals 1.0, and the same is true for C/G. However, the molar concentrations between AT and CG vary with the sequence of DNA analyzed. In addition to the AT and CG pairings, other rare can be found. The Hoogsteen Pairing that occurs in triple strands of RNA and DNA and the pairings GA, CA, and TG are examples of these rare events. The glycosidic bonds in DNA between the deoxyribose and the nitrogenous bases are not directly opposite in the double helix, generating two unequal cavities in its contour. The two cavities are called the greater cavity and the smaller cavity. In these cavities, especially larger ones, the bases are chemically distinguishable and exposed to the solvent medium. Thus molecules (usual proteins) that interact with specific sequences of bases can identify these sequences without breaking the double helix structure. In principle, the strong negative charges of the phosphate clusters from the phosphodiester chains of the two strands of DNA tend to show repulsion. However, a set of forces acts to stabilize the structure of the double helix of the DNA. It is critical that these forces be strong enough to maintain the integrity of the double helix but must allow conformational flexibility, which is essential for its activity. As we shall see, the double helix must be separated during DNA replication and transcription chapters for these vital processes to occur.

Besides the covalent bonds, which bind the atoms in the molecules, other, weaker forces act to stabilize the double helix structure of DNA. These forces are:

1. The hydrophobic effects stabilize the pairing between the bases. The purine and pyrimidine rings, facing the interior of the double helix, are held together by the internal cohesion of water molecules. The hydrophilic sites of the bases are exposed to the solvent in the cavities.
2. Stacking the bases inside the double helix allows the establishment of Van der

Walls forces between the aromatic rings of adjacent bases. These forces are weak but additive in the maintenance of the final structure.
3. The sugar-phosphate chains, which are negatively charged, interact, interact with cations (mainly Mg^{+2}) in solution, neutralizing the repulsion between the two chains and stabilizing the double helix.

In the cell, the DNA molecule is always linked to structural proteins, such as histones or proteins with different functions. These interactions with the proteins also contribute, in a significant way, to the maintenance and changes in the DNA structure. The formation of chromatin is the primary type of structuring of the DNA molecules in the cell. Although DNA is often seen as double-stranded, single-stranded DNA is found in nature, representing viruses' genomes and some bacterial plasmids.

Chemical Properties of DNA

The chemical properties of DNA that are most important for the storage functions of genetic information are the complementarity between the nitrogenous bases of the two chains, the ant-parallelism that gives directionality to chains, and the capacity of denaturation and renaturation of the double helix. Denaturation (or fusion) occurs when the hydrogen bonds between the complementary chains of DNA, which form the double helix, are broken, and the strands get separated. The reverse process, the re-establishment of the bonds between the complementary bases (formation of the double helix), is renaturation (or hybridization). Whenever the molecule of DNA is replicated or transcribed, the DNA strands are denatured and renatured. After the renaturation, all of the double helix's original properties are re-established.

It is possible to denature the DNA *in vitro* in solution by increasing the temperature, titration with acids or alkalis, and denaturing agents, such as formamide, urea, and dimethyl sulfoxide (DMSO), for example. The acids protonate the nitrogenous A, G, and C rings, and the alkalis deprotonate the nitrogenous rings of G and T. These treatments generate charged groups inside the double helix, leading to disruption of the hydrogen bonds between the complementary bases. Since the glycosidic bonds in purines are sensitive at acidic pH, denaturation by acids has little practical application because it causes the purification of the bases, which permanently changes the DNA molecule. The denaturation/renaturation of the DNA can be accompanied by measuring in a spectrophotometer the absorbance of the ultraviolet light. The nitrogenous bases absorb UV light, reaching a maximum wavelength of 260 nm (A260). In the DNA molecule, the nitrogenous bases are responsible for the greatest absorption of UV light, which is maximum when the strands of a double helix of DNA are entirely separated because the bases are entirely exposed to the medium. When the strands of a DNA molecule are entirely separated, the absorption of UV light is 37% higher than that obtained with the DNA in its native form (double helix). This can be demonstrated if a solution of DNA is subjected to an increase in temperature, and the absorbance of the solution is measured at 260 nm. The absorbance increases with temperature until the strands separate completely when the maximum absorbance is recorded (this sudden change is called the hyperchromic effect) and remains constant even as the temperature rises because the bases are already exposed and have maximum absorption of UV light. The temperature at which 50% of the DNA is denatured is called the mean melting temperature (Tm). The reverse effect - the renaturation can also be followed in this way. Thus, if this solution of fully denatured DNA is gradually cooled down, the complementary chains of DNA will re-form the double helix. The absorbance of the UV light will decrease because the nitrogenous bases will be less exposed to the medium and therefore absorb less UV light. The reduction in absorbance ceases when the double helix is entirely renatured (Fig. **3**).

Even when the two DNA strands are entirely separated, the process can be reversed because the complementary strands reassemble. This ringing or hybridization generally occurs at a temperature of 25°C below Tm. In the beginning, this renaturation occurs slowly, but as a few complementary bases are associated, the speed of the process increases dramatically, making this process very fast. Remember that the pairing between the nitrogen-containing strands of DNA occurs by the complementarity of the sequences and, therefore, is dependent on the base sequence of the strands, which makes this process precise. In other words, only two strands that present complementary sequences and are antiparallel (one strand in the 5'→3' direction and the other complementary in the 3'→5' direction) will reassemble. If an abrupt cooling occurs, the DNA strands collapse, and renaturation may not happen. The hydrogen bonds between the bases in DNA do not show the same stability with respect to the denaturing agents due to the presence of two hydrogen bonds between the AT pairs and three bonds between the CG pairs. Therefore, disrupting a CG pair requires higher temperatures, more alkaline pH, or higher concentrations of denaturing agents than separating an AT pair. Thus, the Tm of DNA depends on the proportion of AT in relation to GC (remember that the DNA molecules can have different ratios of AT and GC, depending on the organism or the region of the DNA analyzed). The higher the percentage of GC (index of GC or GC%), the higher the temperature necessary to denature the DNA molecule; therefore, the higher the value of Tm will be. In nature, this can be observed in the genome of some organisms. For example, bacteria that inhabit high-temperature environments tend to have a genome with a higher GC, which would potentially have greater stability at higher temperatures. Another example is regions of the genome where the double helix is separated for the start of replication or transcription that tend to have a higher AT index (lower GC), facilitating the opening (separation of the double helix chains). It is possible to calculate the GC or AT content of a given DNA from the value of its Tm, utilizing the equation:

$$\mathbf{Tm(°C) = 69.3 + 0.41\ (GC\%) - 650/primer\ length}$$

In addition to measuring UV light absorbance, other techniques can be used to monitor the denaturation/renaturation of DNA, for example, sedimentation by centrifugation; enzyme treatment specific for simple strands (nuclease S1, for example); and electron microscopy.

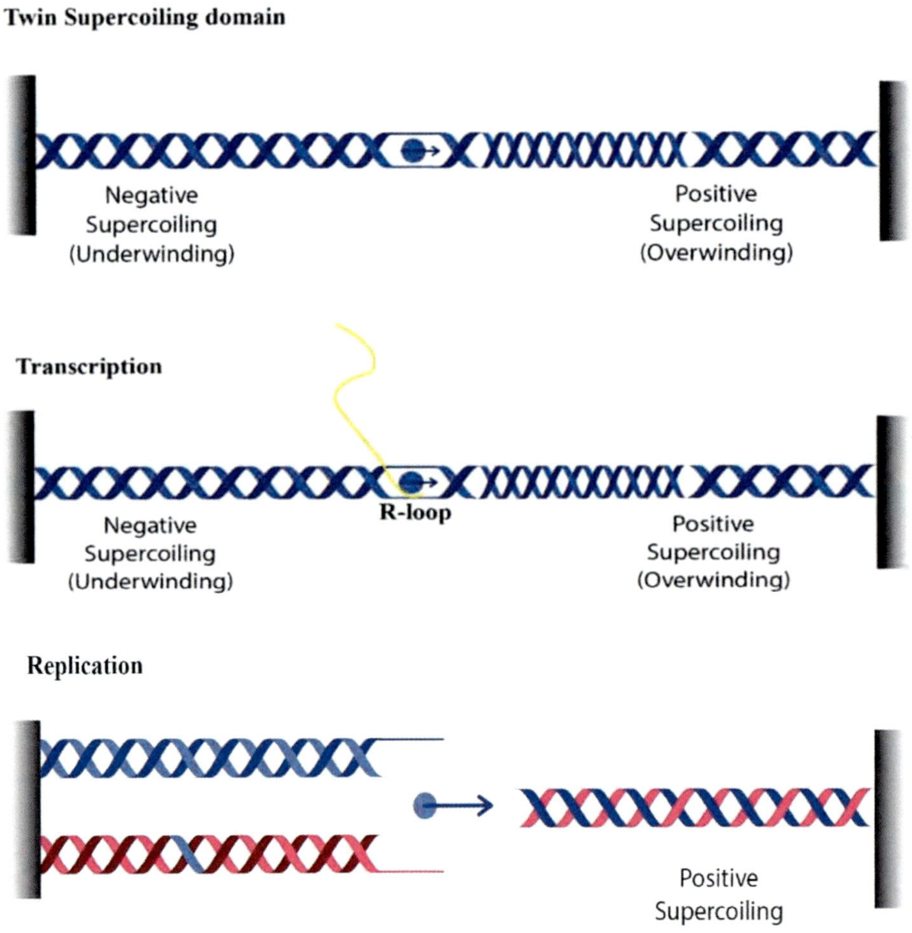

Fig. (3). Supercoling in DNA structure.

Supercoiling in DNA

In addition to the secondary structure of the double helix, DNA assumes a tertiary structure, denominated supercoiled or super-helical. This other structure is basically the coiling of the double helix upon itself. Remember that the DNA molecule is very long, and the supercoiled structure is the predominant conformation of the DNA in the cell and is essential for the packing of the DNA in the genomes and the structure of the nucleosomes. Furthermore, supercoiling is involved in replication, transcription, and recombination, controlling the expression of some genes. The standard methods of DNA isolation alter the molecule's three-dimensional structure. When supercoiling was first described, it was believed to occur only in small, circular molecules, such as viral genomes and plasmids, mitochondrial and chloroplast DNA. It is known that this is a characteristic of almost all DNAs, circular and linear, and is significant for their functionality.

The structures of supercoiled DNA are studied in a branch of mathematics called topology. For these considerations, the double helix DNA molecule can be represented as two elastic bands coiled up, forming the helix-secondary structure. When the ends of the strands are joined, a flat circular structure is formed. As the strands in the double helix are antiparallel, they must be correctly linked, *i.e.*, the 5'-P end is attached to the 3'-OH end. When this structure sits perfectly on a flat surface without changing the geometry of the double helix, it is called relaxed. If, however, before joining the ends, one of the strands is rotated about itself one or more times in 360° turns in the direction of the unwinding of the double helix. In contrast, the other strand remains fixed; the circular molecule formed after the ends are joined will no longer be relaxed. This new situation can be accommodated in three ways:

1. The generated tension can be distributed along the double helix, decreasing the number of times one strand winds around the other. In this way, the number of paired bases around the helix increases, or their stress concentration at a specific point in the circular, which will be unpaired (single strand), maintaining the same number of base pairs around the helix, in both cases, the structure will remain flat.
2. The tension produced can be cancelled out if the helix itself (supercoiling). This way, the number of base pairs around the helix will remain the same, but the structure will no longer be flat but supercoiled in a three-dimensional way. When the double helix unwinding generates this supercoiling, it is called negative supercoiling. This negative supercoiling is found in all organisms. The archaea are exceptions to this rule and have their DNA with positive supercoiling. The supercoiling can assume two types: plectonemic and toroidal

(the spiral type of telephone wires, for example); the first is found in DNA in solution, and the other, in the DNA coiled in proteins such as the histones, which form the nucleosomes.

3. The unwinding of the helices may induce the formation of alternative structures, like cruciform structures, of triplex or Z-type DNA. These alterations can also absorb some of the tension, generating fewer supercoils.

Various methods can measure the degree of supercoiling and its generation and removal in DNA molecules. Agarose gel electrophoresis allows the separation of DNA molecules according to the molecular mass (size), shape, and compaction. The sedimentation speed is another example because DNA molecules with different degrees of coiling sediment with different rates and, therefore, can be separated by centrifugation in gradients. Finally, the supercoiling can also be visualized by electron or atomic force microscopy by directly observing molecules with different degrees of supercoiling.

The degree of supercoiling of DNA in the cell is fundamental for its functionality. The DNA is maintained in a condition of negative supercoiling that supplies energy for the denaturation of the double helix, allowing the access of enzymes that promote the replication, transcription, repair, and recombination of DNA. Current knowledge shows that this supercoiling is maintained by the cell in a homeostatic system and that, in addition to the primary structure of DNA (base sequence), the supercoiling also stores information. This information is essential for its functionality. In addition, the supercoiling allows regions of genomes far apart to be spatially close.

DNA Topoisomerases

To maintain homeostasis of the supercoiling and DNA conformation in the cell, there is a complex system of enzymes, the topoisomerases, acting coordinately. The topoisomerases are enzymes that catalyze the interconversion of DNA topoisomers, allowing alterations in the degree of supercoiling [3]. These enzymes promote the transient breaking of phosphodiester bonds of the DNA strand, generating an intermediate form in which the protein remains covalently linked to the DNA and allows the DNA strands to pass over each other or undergo torsion, thus altering the supercoiling of the molecule. These enzymes introduce or remove supercoils in the DNA and resolve structures, such as knots and concatemers, generated in the cellular processes. All the studied prokaryotic and eukaryotic cells possess topoisomerases, which are fundamental in DNA replication, transcription, recombination, and chromatin remodeling.

Evolution has developed two general strategies for altering DNA topology by topoisomerases. One of them is the rotation mechanism, in which the topoisomerase cleaves the phosphodiester bond of one of the chains of the DNA and allows one of the ends of the open chain to rotate around the helix. Subsequently, the cleaved chain is re-ligated, and the resulting DNA molecule is left with supercoiling different from the initial one (topoisomerases of the type I family). The other strategy involves the cleavage of one or the two chains of DNA of the double helix, causing the breakage of some hydrogen bonds between adjacent bases, followed by the passage of a double helix through the cleaved region (topoisomerases of the type II family). The importance of maintaining the DNA supercoiling for the cell's functionality has led to detailed studies of the topoisomerases since their discovery in 1970. The discovery also boosted these studies that these enzymes are important targets for the action of many substances with applications in medicine. Thus, today, many topoisomerases are known and characterized in different organisms.

In general, there are many subfamilies of the two families of topoisomerases, classified amino acid sequences, and the mechanism involved in the topoisomer formation. The topoisomerases of the type I family are divided into three subfamilies: IA, IB, and IC. The type II family is divided into two subfamilies: IIA and IIB. In all the organisms studied, at least one topoisomerase of each family. The topoisomerases of each organism have received names such as topoisomerase I, gyrase, topoisomerase IV, *etc*., which can create some confusion. Table **1** summarizes the topoisomerases in which organisms are found in each subfamily and some of their activities.

Table 1. Topoisomerases: Short classification (Source, 2002).

Family	Subfamily	Organism/Enzyme	Activity
Type I cleave one strand of DNA	1A	Bacteria and Archaea/Top I and III Eukaryotes/Top III Mimi viruses/Top IA	Undo knots and concatemers Relaxes supercoiling (-) Reverse gyrase of hyperthermophile bacteria induces supercoiling (+)
	1B	Eukaryotes/Top I Viruses and some bacteria	Relax supercoil (-) and (+)
	1C	Only one genus of archaea/topiary V	Relax supercoil (-) and (+)
Type II cleaves the two DNA strands	11A	Eukaryotes, viruses, and bacteriophages/Top II Bacteria/top IV Bacteria, archaea and chloroplast, DNA Gyrase	Relax supercoiling (-) and (+). Undo knots and concatemers Relaxes supercoiling (+). Induces supercoiling (-)
	11B	Archaea, plants, and some algae/Top IV	Relax supercoil (-) and (+) Undo knots and concatemers

Types of DNA

The types of DNA found under physiological conditions are: DNA type B, DNA type A, and DNA type Z. These types have been identified by the analysis of the crystallographic structure and are generated by the composition of bases due to the environment in which they are located, or due to their binding with specific proteins. Alterations in the conformation can facilitate or hinder the interaction of the DNA with the proteins. What differentiates these three types of DNA is their thickness, the number of base pairs per turn of a helix, and the exposure of the nitrogenous bases to the external environment.

DNA type B is the physiological one primarily found in the cell. DNA type A occurs in very low humidity conditions and when mixed strands of DNA--RNA and RNA double strands are formed. DNA-RNA hybrids are created in transcription, and double strands of RNA are included in the genomes of some viruses and, in the cell, in the transfer of RNAs and ribosomal RNAs. DNA type Z occurs in short regions of the DNA, where sequences of cytosines and guanines exist. DNA type B presents the form of the classical double helix described by Watson and Crick. This type of DNA occurs in high relative humidity (92%) and solutions of low ionic strength. It presents the following characteristics:

1. The double helix turns right, and the rotation between the two adjacent base pairs is 34.6°. Thus, the double helix makes a complete turn every 10.4 bp.
2. One turn of the helix travels a distance of 3.40 nm, which is the pitch of the helix. Each base pair adds to the helix a distance of 0.33 nm.
3. The diameter of the double helix is 2.37 nm.

If the type B DNA in solution is dehydrated, for example, in 75% ethanol, or if the salt content is decreased, the molecule, which is thin and long, will shorten and thicken, generating the DNA type A. The larger cavity will become narrower and deeper, and the smaller cavity will become broader and shallower. The pitch of the helix reduces from 3.40 nm to 2.60 nm per turn, and the number of base pairs per turn increases to 11. In both type B and type, A DNA, the deoxyribose and the nitrogenous base are on opposite sides of the glycosidic bond, that is to say, in the anti-conformation. This situation minimizes the steric repulsion between the base and the sugar. In the presence of high concentrations of cations, some nucleotides assume the syn-conformation, with the sugar and the base on the same side of the glycosidic bond. Under these conditions, a different structure, the DNA type Z, is formed. The double helix in this type of DNA is coiled to the left and switches the conformations syn and anti, typical in sequences where the nucleotides G and C alternate themselves. The pyrimidines conserve the standard anti-conformation, with the sugar in the C2'-endo form and purines in syn and

C3'-endo form. The double helix turns to the left under these conditions due to the zigzag appearance of the chain, this shape, described in synthetic oligonucleotides, has been denominated Z, and type Z is longer and thinner than type B. A complete turn of the helix occurs every 12 bp, and the pitch of the helix increases to 4.56 nm, with a diameter of 1.84 nm. The larger cavity disappears, and the smaller one becomes very deep, forming a spiral around the structure. Some regions of the DNA *in vivo* can be in this form.

Some factors stabilize the formation of the DNA type Z, such as the methylation or the bromination of the bases and the torsional stress, as occurs in regions of the DNA supercoiled negatively or when specific proteins are connected to the DNA. Two factors determine the possibility of the occurrence of type Z DNA regions: the sequence of nucleotides and the overall structure of the double helix. If these are propitious, the DNA can convert from type B to type Z under natural conditions. This seems to occur in regions of the genome that are being actively transcribed. The Z-type DNA may be necessary for regulating gene expression and involved in DNA processing and genetic instability events.

Although the double helix is the predominant form of the DNA in nature, molecules of single-stranded DNA make the genomes of some animal and vegetable viruses and those of some bacteriophages, which are bacterial viruses. Some bacteriophages that infect the bacterium *E. coli*, for example, the bacteriophages M13 and X174, show single-strand DNA as a genome. This DNA produces a replicative form (double-strand DNA), which is the template for synthesizing the DNA of the bacteriophages replicated in the host.

Other DNA Structures: Curvature

Long DNA strands are very flexible; however, sequences smaller than 100 bp are relatively rigid to be curved. Some specific sequences and particular proteins make certain DNA regions more susceptible to bending. These regions, which are more flexible to bending, are essential in sequences that control replication, transcription, and recombination processes. DNA sequences constituted by parts of adenines have their curvature facilitated, principally if these sequences are interspersed and repeat themselves various times. The binding of proteins may also facilitate the curvature; a classic example is a nucleosome, in which a segment of DNA of 145 bp coils almost twice around the octamer. The cavities in contact with the protein should narrow, and the wider opposite cavities should be more expansive. Many other proteins, besides histones, promote bend in the DNA: transcription factors, enzymes of recombination, and enzymes that modify the DNA, among others. These proteins bend the DNA with different magnitudes. There are various functions for the bending of DNA, such as:

- Condense and package DNA, as in nucleosomes.
- Expose the sequence of bases internalized in the double helix.
- Approximate distant binding sites in the linear DNA.
- Forming unique protein-DNA structures to align specific sites, as in the case of site-specific recombination.
- Forcing the DNA molecule to stimulate cleavage or denaturation.

Cruciform Structures

There are sequences in DNA that show symmetry, containing repeated and inverted regions. For example:

ATTCGCGTAGTAGACATAGCTGACATAGTCAGCTATGTCTGCTC

Sequences like these can form structures where the pairing between the two strands is replaced by the pairing between complementary bases in the same strand.

ATTCGCGTAGTAGACATAGCTGACA

::::::::::::::::::::T

CTCGTCTGTATCGACTGA

This change reduces the number of turns in the double helix region, removing the degree of negative supercoiling by one unit for every ten pairing bases. These structures occur *in vivo* and are fundamental for specific processes.

Holliday Junctions

Holliday junctions are structures formed by four DNA strands, essential intermediates in the recombination. Specific proteins and enzymes recognize these structures and determine how they will be cleaved, giving rise to the recombination products.

RIBONUCLEIC ACID (RNA)

One of the molecules that have generated new knowledge in molecular biology in the last decade has been RNA [4]. Discovered at the same time as DNA, the studies with RNA had, at first, a slower development. The number of different classes of RNA and its involvement in innumerable, previously unknown processes in the cell was the main responsible for the delay in their characterization and the discovery of their functions. The number of biological functions in which RNAs are involved is increasing. Approximately ten years ago,

RNAs were attributed functions, primarily in synthesizing proteins. The messenger RNA (mRNA) was seen as an intermediary molecule in transferring information in the genome to synthesize proteins. Its conformation appeared to be random and unimportant. The conformation of ribosomal RNA (rRNA), generated by its secondary and tertiary structures, seemed to be a mere framework for interactions with ribosomal proteins, catalyzing the synthesis of proteins. The transfer RNAs (tRNAs) were studied more in terms of tertiary structure but with a view to their ability to carry the amino acids and interact with ribosomes. Nowadays, RNAs and their spatial conformation are extensively studied. Several new classes of RNA have been discovered that are involved in controlling gene expression. There are new possibilities for interactions of RNAs with proteins, DNA, and other RNAs, from a diversity of spatial conformations. As to the possible biochemical activities of the RNAs, the discovery that they may catalyze reactions led to the formulation of the currently called "the world of RNA," a molecule of linear RNA that can, in its sequence of bases, store genetic information (similar to DNA); perform an internal pairing, forming a secondary structure, exposing or hiding specific sequences and, thus, interact with other molecules; it can assume a spatial conformation, tertiary structure, structure, containing surfaces that can interact with other molecules or internal regions that can create metal-binding and promote catalysis, such as enzymes. This malleability of function is not exhibited by any other macromolecule in the cell, thus allowing us to propose that RNA has been the primordial molecule and evolved for even more functions, such as regulating gene expression. Therefore, the knowledge of the structure and conformation of RNA is even more relevant.

Chemical Composition of RNA

There are similarities between the structure of RNA and that of DNA; both are linear polymers of subunits of nucleotides linked to each other by phosphodiester bonds 5'→3'. However, the sugar is ribose in the RNA molecule, and the nitrogenous base thymine (T) is replaced by the uracil (U). Therefore, RNA comprises adenine, cytosine, guanine, and uracil. The presence of ribose gives RNA its name of ribonucleic acid. RNA is predominantly single-stranded but forms regions of double-strand by internal pairing. These regions vary significantly in length, depending on the class of RNA and its function (Fig. **4**).

RNA is more reactive than DNA; the subtle difference in the presence of the hydroxyl at the 2' position of the ribose has important reflections on the structure and reactivity of RNA. Thus, the DNA is stable in the presence of alkali because, even though the strands are denatured with the breaking of the hydrogen bonds, the phosphodiester bonds in the chains remain unaltered. In the presence of alkali, RNA is hydrolyzed, yielding a mixture of 2' and 3' monophosphate nucleosides.

In the presence of hydroxyl ions, the 2'-hydroxyl group of ribose is converted to the alkoxy anion. This potent nucleophilic agent reacts with the phosphodiester, forming a cyclic nucleotide, 2',3'-nucleoside monophosphate. In this process, the 5'→3' phosphodiester bond is broken. Next, the cyclic nucleotide is subjected to the second hydrolysis, in which a hydroxyl ion breaks the phosphodiester cycle, forming a mixture of 2' and 3' monophosphate nucleosides [5].

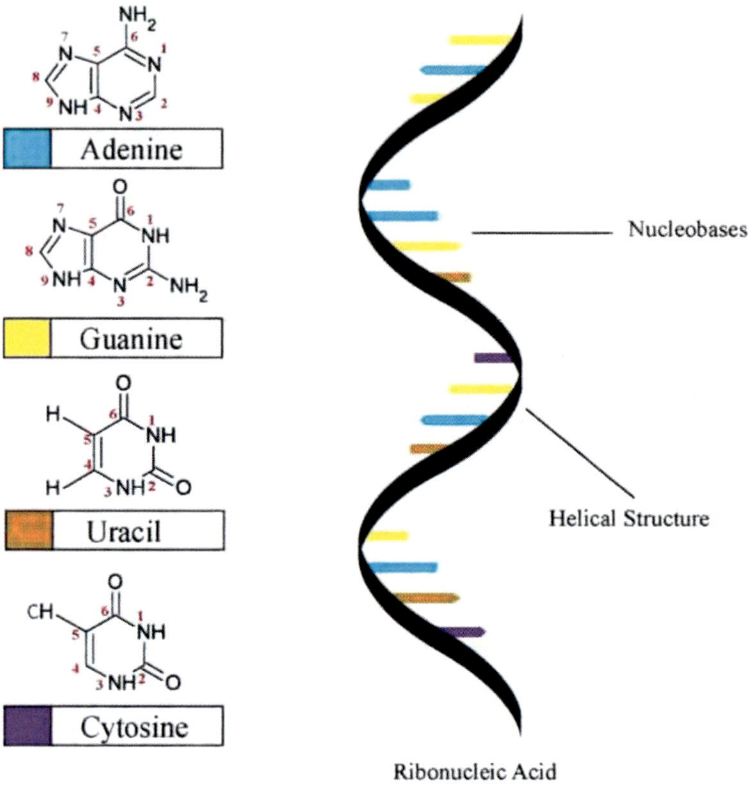

Fig. (4). Structure of RNA.

Secondary Structure

Pairings between CG and AU occur between complementary regions of the same RNA molecule or, less frequently, between distinct molecules. The presence of the hydroxyl at carbon 2' of the ribose also causes significant spatial differences concerning the position of the phosphate group relative to the nitrogenous base. In RNA, double helix formation also occurs in regions with the pairing between two sequences. In this case, its structure resembles DNA type A. The structure of the DNA type B is not being formed in the RNA due to the spatial impediment represented by the hydroxyl grouping in the 2' position of the ribose. Hybrids RNA-DNA are developed in different processes in the cell, for example, in transcription, and their structure is type A. The secondary and tertiary structures of RNA are essential in its functions and the recognition of protein-RNA recognition.

Classification of RNA

In contrast to DNA, different types of RNA are present in cells and have specific functions. The RNAs are classified according to their location and function in the cell. Advances in recent years, mainly in genome characterization and high-performance methods for evaluating transcription, have led to several new classes of RNAs. Some of these classes still have functions that are not fully understood, others are not yet fully understood, and others are likely to be described. Therefore, it is important to present a general classification of the RNAs. The main types of RNA found in cells directly involved in protein synthesis are the following:

1. mRNA (messenger RNA) transfers genetic information from DNA to ribosomes, where protein synthesis occurs. It represents 1 to 5% of the total RNA of the cell.
2. rRNA (ribosomal RNA) is the major component of the ribosomes. It represents about 75% of the total RNA of the cell, forming double strands by internal pairings.
3. tRNA (transfer RNA) transports residues from amino acids to the ribosomes to synthesize proteins. It represents 10 to 15% of the total RNA of the cell.

The discovery of the other classes of RNA led to a broader subdivision of RNAs into coding and non-coding RNAs. Coding RNAs are exclusively those containing genetic information for synthesizing proteins. They are the mRNAs of prokaryotic and eukaryotic cells. In the case of eukaryotic cells, mRNAs are produced, at first, as precursors called pre-mRNAs, which go through various processes to make the final active mRNA. All other RNAs are called non-coding RNAs and include rRNA and tRNA. These classes (mRNA, rRNA, and tRNA) participate directly in protein synthesis and are called RNAs of primary cellular metabolism (housekeeping).

In eukaryotic cells, besides these maintenance RNAs, the organelles, the mitochondria, and chloroplasts have their mRNAs, rRNAs, and tRNAs with characteristics similar to prokaryotic RNAs to carry out the synthesis of proteins from the genomes of the organelles. In prokaryotes, there are two more classes of non-coding RNAs, the sRNAs (from small) and the CRISP RNAs (from clustered regulatory interspaced short palindromic repeats). The sRNAs are small (between 100 and 300 nt), can pair with specific mRNAs, bind to proteins, and regulate gene expression by altering the mRNA's stability or modulating translation. The

CRISP RNAs are sequence homologous to the genomes of bacteriophages (bacterial viruses) and plasmids and interfere with bacteriophage infection and plasmid conjugation. In eukaryotic organisms, in addition to non-coding RNAs (rRNA and tRNA) are also found to be small nuclear RNAs, snRNAs (from small nuclear), the small cytoplasmic RNAs, scRNAs (from small cytoplasmic), and the nucleolar RNAs (from the nucleolus), snoRNAs (small nucleolar). These RNAs are also involved in the basics of the cell's metabolism. The snRNAs are found in the nucleus, forming complexes with proteins and ribonucleoproteins (RNPs), and participate in removing introns. The snRNAs are located in the nucleolus (region of the nucleus where rRNA synthesis takes place) and participate in the processing of ribosomes.

The other classes of eukaryotic RNA are involved in regulating gene expression. They are non-ncRNA (non-coding) RNAs divided into short and long groups. The short ncRNAs are:

- The microRNAs.
- miRNAs (from micro).
- The small interference RNAs.
- siRNA (small interfering, sometimes only iRNA).
- piRNAs.

The miRNAs and siRNAs are found in all eukaryotic cells. piRNAs are found only in germ cells of animals. The long ncRNAs, lncRNA (from long non-coding), are found both in the nucleus and in the cytoplasm and are usually transcribed and have regulatory functions that are not yet fully understood.

GLOSSARY

DNA: Deoxyribonucleic acid is a polymer composed of two polynucleotide chains that coil around each other to form a double helix carrying genetic instructions for the development, functioning, growth, and reproduction of all known organisms and many viruses.

RNA: Ribonucleic acid (RNA) is a polymeric molecule essential in various biological roles in the coding, decoding, regulating, and expressing of genes. RNA is assembled as a chain of nucleotides, but unlike DNA, RNA is found in nature as a single strand folded onto itself rather than a paired double strand.

Purines and Pyrimidines: Purines and pyrimidines are the nitrogen bases that hold DNA strands together through hydrogen bonds. They pair together through complementary pairing based on Chargaff's Rule (A: T and G: C). The purines in

DNA are adenine and guanine, the same as in RNA. The pyrimidines in DNA are cytosine and thymine; in RNA, they are cytosine and uracil.

Gene Expression: It is the process by which information from a gene is used to synthesize a functional gene product that enables it to produce end products. These products are often proteins.

REVIEW QUESTIONS

1. The set of RNA molecules present within a cell changes over time. Explain the cause.
2. DNA and RNA are molecules with a large amount of information. Explain the significance and implications of this statement.
3. It has been suggested that DNA from fossil remains could be extracted, cloned, and used to revive extinct species. Comment on the feasibility of this idea.

REFERENCES

[1] Brosnan CA, Voinnet O. The long and the short of noncoding RNAs. Curr Opin Cell Biol 2009; 21(3): 416-25.
[http://dx.doi.org/10.1016/j.ceb.2009.04.001] [PMID: 19447594]

[2] a. Carninci P. The long and short of RNAs. Nature 2009; 457(7232): 974-5. b. Carthew RW, Sontheimer EJ. Origins and mechanisms of miRNAs and siRNAs. Cell 2009; 136(4): 642-55.
[PMID: 19239886]

[3] a. Champoux JJ. DNA topoisomerases: Structure, function, and mechanism. Ann Rev Biochem 2001; 70(1): 369-413. b. Cooper TA, Wan L, Dreyfuss G. RNA, and disease. Cell 2009; 136(4): 777-93.

[4] Draper DE. RNA folding: Thermodynamic and molecular descriptions of the roles of ions. Biophys J 2008; 95(12): 5489-95.
[http://dx.doi.org/10.1529/biophysj.108.131813] [PMID: 18835912]

[5] Martin J. Describing the structural diversity within an RNA's ensemble. Entropy 2014; 16(3): 1331-48.
[http://dx.doi.org/10.3390/e16031331]

CHAPTER 3

DNA Replication

Kainat Gul[1,*]

[1] *Department of Microbiology, Abbottabad University of Science and Technology, Havelian, Abbottabad, Pakistan*

Abstract: DNA replication is the process by which a cell makes identical copies of its genetic material. This process is essential for transmitting genetic information from one generation of cells to the next during cell division. DNA replication is a complex process involving unwinding the double helix, separating the two strands, and synthesizing new complementary strands. The process is catalyzed by a large complex of proteins called the replisome, which includes DNA polymerases, helicases, and other enzymes. The accuracy of DNA replication is critical to the maintenance of genetic information, and errors in replication can lead to mutations and genetic diseases. Understanding the molecular mechanisms of DNA replication is fundamental to many fields of biology, including genetics, biotechnology, and medicine.

Keywords: Biotechnology, Cell division, Complementary strands, DNA polymerase, Genetic information, Helicase, Mutations, Molecular mechanisms, Replisome.

INTRODUCTION

Cell division occurs in all organisms and is usually described as a sequence of events that result in genome duplication, called DNA replication. After cell division, the same genetic material must be present in all the cells in both eukaryotic and prokaryotic organisms. For cells to maintain the same genetic material, the entire genome must be precisely replicated before each cell division event. DNA replication requires several enzymatic activities that must act in an integrated and sequential manner in the initiation, chain elongation (or growth), and termination stages. The set of proteins that function in replication is called a replisome. As a fundamental principle of repetition, it can be noted that it will only occur if there is a DNA molecule to be copied (template strand), with a region paired containing a free 3'-OH end where synthesis of the new strand of DNA will begin. The DNA molecule to be replicated can be either double-

* **Corresponding author Kainat Gul:** Department of Microbiology, Abbottabad University of Science and Technology, Havelian, Abbottabad, Pakistan; E-mail: kainatgul195@gmail.com

stranded or single-stranded. In instances where specific viral genotypes require the replication of DNA with a single strand, it is imperative to have a double-stranded intermediate before initiating the replication process. In addition to these characteristics, some basic rules for the replication process can be applied to the genomes of all prokaryotic and eukaryotic organisms. The addition of nucleotides is always in the direction of 5'-P to 3'-OH (5'→3'); replication always starts at specific origins.

Creating new DNA strands complementary to the original strands occurs during the cell cycle's S phase. Before cell division, a cell repeats its genetic material through DNA replication. This procedure is crucial for retaining the genome's genetic data and ensuring it is appropriately transmitted to daughter cells. Several proteins and enzymes work in unison to replicate DNA, which is a tightly controlled process. The DNA double helix is unwound during initiation when a collection of proteins binds to particular locations on the DNA, known as replication origins. After the DNA has been unraveled, DNA polymerases can start the process of elongation, which involves the synthesis of new DNA strands. Finalization marks the completion of DNA replication, after which the freshly created DNA strands are examined for mistakes and bundled into chromatin in preparation for cell division. All living things must accurately and effectively replicate their genetic material to survive and function. Maintaining the integrity of the genome and preventing the onset of illnesses like cancer require the management of DNA replication. The correct and effective replication of genetic material is regulated by replication origin firing, replication fork stability, replication termination, replication stress response, replication fidelity, replication licensing, and replication-coupled chromatin assembly. The cell can ensure that accurate genetic information is accurately transmitted from one generation to the next and that the genome remains stable and functional by strictly controlling DNA replication.

In addition to regulating DNA replication, the replication process can be affected by various factors, including DNA damage, mutations in replication-related genes, and disruptions to the DNA replication machinery. When DNA damage occurs, cells activate a complex network of DNA damage response pathways that can repair or trigger cell death if the damage is too severe. Mutations in replication-related genes, such as DNA polymerases or helicases, can lead to errors in DNA replication and result in genetic instability or diseases such as cancer.

In addition, it is worth noting that disturbances in the DNA replication apparatus, such as the stalling or collapse of replication forks, can give rise to genomic instability and pathological conditions. The phenomenon of replication fork stalling arises when the progression of the replication fork is impeded by an obstruction, such as a DNA lesion or a protein complex, which hinders the synthesis of DNA. If the replication fork cannot be restarted, it can collapse, leading to the loss of genetic material and chromosome instability. Understanding the regulation and factors affecting DNA replication is essential for developing new treatments and therapies for diseases associated with genetic instability, such as cancer [1, 2].

UNIDIRECTIONAL AND BIDIRECTIONAL REPLICATION

The replication starts at the origin and proceeds sequentially along the DNA strand in one or both directions to termination, forming a replication bubble. In this replication bubble lies the region of DNA that has already been replicated and flanked by regions where new strands have not yet been synthesized. A replication fork (replisome) starts from the origin in unidirectional replication and continues replicating DNA in one direction, whereas in bidirectional, two replication forks leave the head in opposite directions [3, 4].

In 1958, Meselson and Stahl's experiments demonstrated that DNA replication occurs semiconservative, with each strand of DNA copied into the direction 5'→3', originating new strands. In this replication, the two strands of parental DNA must be separated, and each one will serve as a template for synthesizing new DNA strands. This process is considered semiconservative, as the DNA molecules generated after replication contain only one newly synthesized strand, and the other corresponds to the parental DNA strand. One strand is oriented in the direction 5'→3', and the other in the 3'→5' order (anti-parallelism). Each newly synthesized strand is complementary (AT pairing and GC) to the template strand and presents antiparallel conformation.

Initiation

The first stage of DNA replication is the initiation, in which the replication machinery is assembled at specific sites on the DNA molecule called replication origins. The initiation of DNA replication is a complex process that involves regulatory mechanisms and several proteins. It begins with recognizing the image's origin by a protein complex called the origin recognition complex (ORC) in eukaryotes and the DNAA protein in bacteria. ORC and DnaA bind to specific DNA sequences within the origin, which are rich in adenine-thymine (AT) base pairs and are referred to as AT-rich regions. These regions are easy to denature, which is required for DNA replication. After binding to the origin, ORC and DNA recruit other proteins to form a pre-replication complex (pre-RC) in eukaryotes and a pre-initiation complex (pre-IC) in bacteria. These complexes comprise several proteins, including helicases and single-stranded DNA-binding proteins.

Helicases are enzymes that unwind the double-stranded DNA molecule, creating a replication fork with two single-stranded DNA templates.

The next step in initiation is loading the replicative DNA polymerase onto the DNA template. A protein complex accomplishes this called the replicative DNA helicase that binds to the pre-RC/pre-IC and moves along the DNA molecule, separating the two strands. During the process of helicase movement, it generates regions of DNA with a single strand that are subsequently coated with single-strand DNA-binding proteins. This coating serves the purpose of preventing the re-annealing of the separated DNA strands.

Lastly, the DNA template is loaded with the replicative DNA polymerase, and DNA synthesis starts at the replication fork. DNA polymerase uses the complementary strand to guide the addition of nucleotides to the expanding DNA strand. DNA polymerase then extends the new strand in the 5' to 3' direction as the helicase continuously unwinds the double helix along the DNA. Mutations in replication origin sequences, variations in the activity of initiation proteins, and changes in the amounts of checkpoint proteins are only a few factors that might affect the initiation of DNA replication. For instance, mutations in the DNA protein can cause replication initiation abnormalities in bacteria and have been associated with cancer.

In eukaryotic cells, the initiation of DNA replication occurs at multiple replication origins, allowing for the rapid synthesis of large amounts of DNA. In contrast, bacteria have a single source of image, which is sufficient for replicating their small genome. The initiation of DNA replication is a critical step in the cell cycle and must be tightly regulated to prevent errors and maintain genomic stability. Defects in the initiation process can lead to DNA damage, chromosomal abnormalities, and cancer.

Elongation

The elongation process in DNA replication refers to the subsequent stage following initiation, wherein the replication machinery effectively generates fresh DNA strands by utilizing the single-stranded templates formed during the initiation stage. The elongation process encompasses multiple proteins, such as the replicative DNA polymerase and various accessory proteins, stabilizing the DNA molecule and synchronizing the replication fork's movement. During the elongation process, the DNA polymerase responsible for replication incorporates nucleotides into the developing DNA strands in a direction from the 5' end to the 3' end. This occurs by utilizing the complementary strand of DNA as a template. The polymerase enzyme is responsible for deciphering the nucleotide sequence of the template strand and subsequently incorporating nucleotides that are complementary to the template sequence (Fig. **1**). During DNA replication, the polymerase enzyme traverses the template strand, synthesizing a novel strand of DNA that is complementary in sequence.

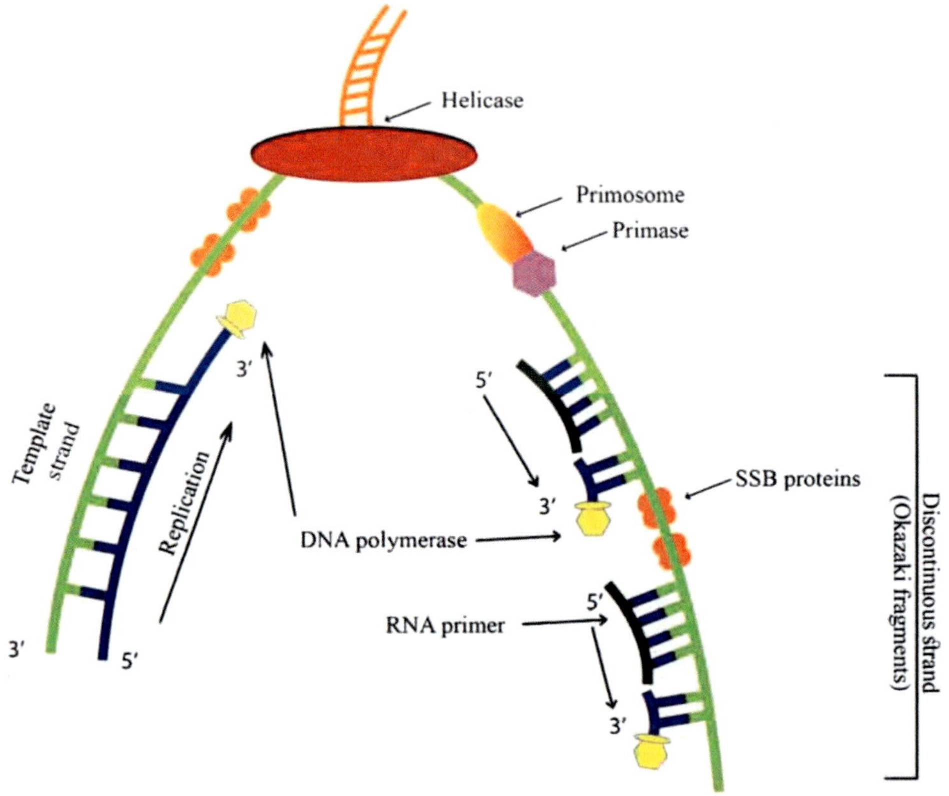

Fig. (1). Start of DNA replication. Continuous and discontinuous chains are observed; both replicate in the direction from 5' to 3'.

After the replication bubble's formation, DNA polymerases synthesize fresh DNA strands, proceeding in a 5' to 3' orientation. The process of DNA replication entails the synthesis of two complementary strands, namely the leading strand and the lagging strand. The DNA replication method involves synthesizing two leading and lagging strands. Okazaki fragments synthesize the lagging strand discontinuously, while the top strand is generated constantly. The enzyme primase is responsible for the synthesis of RNA primers, and the DNA polymerase recruits it. The DNA polymerase employs these primers as a starting point for DNA synthesis. The elongation process is supported by several auxiliary proteins that stabilize the DNA molecule and coordinate the replication fork's movement. The involvement of helicases and single-stranded DNA binding proteins is of utmost importance in the process of DNA unwinding and preventing the recombination of single-stranded templates. Topoisomerases represent a category of enzymes that are pivotal in managing the buildup of supercoiling within the DNA structure. The mechanism of DNA strand fragmentation and subsequent recombination accomplishes this phenomenon.

DNA synthesis exhibits a high degree of coordination, wherein numerous proteins play crucial roles in ensuring the accurate synthesis of DNA strands. The proteins in this category consist of single-stranded binding proteins, which impede the recombination of DNA strands, and topoisomerases, which alleviate the accumulated tension preceding the replication fork. DNA replication initiation is tightly controlled and plays a crucial role in ensuring the accuracy and efficiency of DNA synthesis. Apart from the proteins mentioned above, various additional factors participate in the initiation process, including origin licensing and checkpoint proteins. Proteins involved in origin licensing ensure the strict regulation of replication origins, allowing their utilization only once during each cellular cycle. During the G1 phase of the cell cycle, certain factors interact with the replication origins, thereby inhibiting the untimely initiation of DNA replication. The origin licensing factors are eliminated upon entering the S phase, enabling the initiation process. Checkpoint proteins play a crucial role in the initiation process, facilitating the smooth progression of DNA replication. The proteins in question serve as surveillance mechanisms during DNA synthesis, enabling them to impede the process in the event of error detection or when the replication fork encounters obstructions.

Termination

DNA replication continues until the replication forks from opposite directions meet each other. The two newly synthesized DNA molecules are separated, and the replication process is complete. The final step in DNA replication is the proofreading and repair of any errors that may have occurred during the process. Termination is the final stage of DNA replication, during which the replication machinery completes the synthesis of the new DNA strands and disengages from the template DNA. The termination process is relatively simple in prokaryotes but more complex in eukaryotes.

When two replication forks of a prokaryotic cell meet at what is known as the termination site (sometimes referred to as the Ter site) on the DNA molecule, the replication process comes to an end. A specific sequence of DNA may be found at the Ter site, and this sequence binds to a termination protein called Tus. Tus acts as a barrier that prevents the replication machinery from continuing to synthesize

DNA once it has reached the Ter site. The two new DNA molecules are separated once the divisions have converged, and the DNA synthesis is complete, DNA replication termination is more complex and involves several steps. The replication forks meet at sites called replication termination zones, characterized by accumulating particular proteins. The termination process is coordinated by several factors, including the protein complex called the cohesin complex, which helps to hold the newly synthesized sister chromatids together until the termination process is complete. Once the replication forks have converged at the termination zone, the newly synthesized DNA strands are separated from the template strands by a specialized protein complex called the replisome disassembly complex. The replisome disassembly complex breaks down the replication machinery and releases it from the DNA molecule, allowing the newly synthesized DNA to be packaged into chromatin (Fig. **2**).

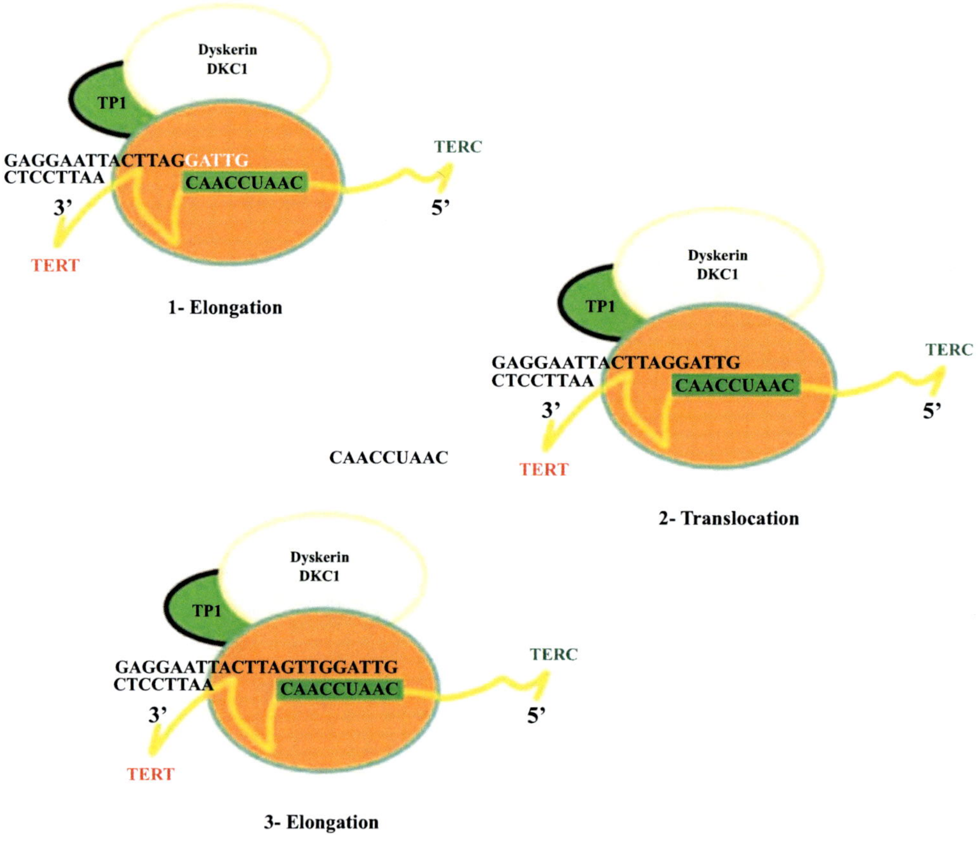

Fig. (2). Telomerase functions in its three phases: elongation, translocation, and elongation.

In eukaryotes, DNA replication termination is tightly regulated to ensure that replication occurs only once per cell cycle. Failure to regulate DNA replication termination can result in genomic instability and contribute to cancer development. One mechanism that helps ensure proper DNA replication termination in eukaryotes is the activation of the replication checkpoint, which monitors DNA replication and ensures that the process proceeds promptly and accurately. If DNA replication is not completed or errors occur, the replication checkpoint can delay cell division until the errors are corrected or the replication process is completed. Another mechanism that helps to regulate DNA replication termination in eukaryotes is the activity of telomeres, specialized structures that caps ends of linear chromosomes. Telomeres help to protect the ends of chromosomes from degradation and fusion with other chromosomes. The telomeres are shortened during DNA replication, and cell division can no longer occur when they become too short. This acts as a natural barrier to cell proliferation and helps to prevent the accumulation of mutations that can lead to cancer.

PROTEINS INVOLVED IN DNA REPLICATION

DNA replication is a complicated process that depends on how several proteins work together. Together, these proteins unwind the double helix of DNA, make new strands of DNA, and check the unique DNA for mistakes. Here are a few proteins that play a role in DNA replication [5, 6]:

DNA Helicase: DNA helicase is a motor protein that unwinds DNA's double helix by dissolving the hydrogen bonds between the strands of DNA that make up the molecule. This results in the formation of a replication fork, at which point both DNA strands can be copied independently.

DNA Polymerase: DNA polymerase, the enzyme that makes new DNA, is the second most crucial enzyme in the cell. DNA polymerases come in various forms, each serving a particular purpose. DNA polymerase-I is responsible for eliminating RNA primers and filling in gaps in the DNA. At the same time, DNA polymerase-III is the primary enzyme responsible for duplicating the leading and lagging DNA strands.

Primase: Primers are small RNAs synthesized on the lagging DNA strand by an RNA polymerase called primase. DNA polymerase-III will use these primers as a jumping-off point when creating a new DNA strand.

Single-Strand Binding Proteins: During DNA replication, single-stranded DNA is produced; single-strand binding proteins bind to this DNA and keep it from re-annealing. The DNA polymerase can then use this strand as a guide to create a new strand of DNA.

DNA Ligase: DNA ligase is an enzyme that joins the short DNA sections generated on the lagging strand during the synthesis process. This is achieved by developing phosphodiester linkages between the nearby nucleotides.

DNA Topoisomerase: DNA topoisomerase is a protein that breaks apart and reassembles DNA strands, relieving tension on the DNA that builds up before the replication fork. This permits the strands to spin around one another, avoiding a tangled mess of DNA.

DNA Proofreading Proteins: DNA polymerase-III is an example of a DNA proofreading protein that may detect and repair mistakes made during the replication of DNA. This makes a 100% accurate replica of the original DNA from which the new DNA was created.

RFC Complex: This protein complex consists of several subunits and loads DNA polymerase onto the template strand. During replication, it aids in maintaining the integrity of the DNA polymerase-DNA complex.

PCNA: Proliferating cell nuclear antigen (PCNA) is a sliding clamp protein that tethers DNA polymerase to the DNA strand, increasing the processivity of DNA polymerase. PCNA also aids in bringing other proteins, like DNA ligase and DNA proofreading proteins, to the replication fork.

RPA: Replication Protein A (RPA) binds to already double-stranded DNA, protecting it from degradation by nucleases. RPA also aids in bringing in DNA polymerase and DNA ligase to the replication fork.

MCM Complex: The DNA replication maintenance complex is a protein complex with many subunits that aid in unwinding the DNA double helix in preparation for the replication fork. By utilizing ATP to sever hydrogen bonds in DNA base pairs, it performs the function of a DNA helicase.

Telomerase: Telomerase is an enzyme that extends the ends of linear chromosomes with repeating DNA sequences called telomeres. This aids in ensuring that no information is lost as DNA replicates.

REGULATION OF DNA REPLICATION

Multiple variables contribute to the intricate regulation of DNA replication. To prevent errors in the transmission of genes, DNA replication must be timed to coincide with other cellular activities, including cell division and DNA damage repair.

One of the key regulatory factors involved in DNA replication is the cyclin-dependent kinase (CDK) protein complex. CDKs are enzymes that control the progression of the cell cycle, and they play a critical role in regulating the initiation of DNA replication. CDKs phosphorylate several proteins involved in DNA replication, including the ORC and the replicative helicase, which helps to ensure that DNA replication occurs at the appropriate time during cell cycle. In addition to CDKs, other factors, such as the availability of nucleotides and activity of DNA damage response pathways, can also affect the regulation of DNA replication [7, 8].

The regulation of DNA replication is a critical process that ensures the accurate and efficient replication of genetic material. Several factors, including the availability of nucleotides, the activity of replication initiation factors, and the activity of DNA damage response pathways, and govern the regulation of DNA replication. Here is a detailed note on the regulation of DNA replication:

Availability of nucleotides: The availability of nucleotides is critical in regulating DNA replication. Nucleotides are the building blocks of DNA and are required for the synthesis of new DNA strands. Several factors, including the activity of nucleotide biosynthesis pathways and the availability of extracellular nucleotides, regulate the availability of nucleotides. DNA replication may be slowed or stopped if there is a shortage of nucleotides.

Replication initiation factors: The initiation of DNA replication is a complex process that involves the coordinated action of several initiation factors. These initiation factors are responsible for unwinding the DNA double helix, recruiting DNA polymerase to the replication fork, and coordinating the synthesis of new DNA strands. Several mechanisms regulate these initiation factors' activity, including post-translational modifications, protein-protein interactions, and gene expression changes.

Checkpoints: DNA damage response pathways are critical in regulating DNA replication. Checkpoints are signalling pathways that monitor the integrity of the DNA and ensure that DNA replication proceeds only when the DNA is intact. If the DNA is damaged, checkpoints can halt DNA replication to allow for DNA repair. Several factors regulate checkpoints, including DNA damage response proteins' activity and nucleotide availability.

Epigenetic regulation: Epigenetic modifications, such as DNA methylation and histone modifications, are critical in regulating DNA replication. These modifications can influence the accessibility of DNA to replication factors, which can affect the efficiency and accuracy of DNA replication. Aberrant epigenetic changes have been implicated in several diseases, including cancer.

Replication timing: The timing of DNA replication is also regulated. Different regions of the genome replicate at other times during the cell cycle. The regulation of replication timing is thought to be influenced by several factors, including chromatin structure and the availability of replication factors.

Replication origin firing: The regulation of DNA replication also involves the control of replication origin firing. Replication origins are specific sites on the DNA where DNA replication initiates. The timing and frequency of origin firing are tightly regulated to ensure that DNA replication proceeds smoothly and efficiently. Several factors, including chromatin structure, the availability of replication factors, and the activity of checkpoint pathways, influence the regulation of replication origin firing.

Replication fork stability: The stability of the replication fork is critical for ensuring the accuracy and efficiency of DNA replication. Several factors, including DNA damage, collisions with other cellular structures, and changes in the nucleotide pool, can destabilize the replication fork. To maintain the stability of the replication fork, several proteins are involved in stabilizing the fork structure and preventing fork collapse.

Replication termination: The regulation of replication termination is critical for ensuring that DNA replication is complete and accurate. DNA replication ends by coordinating several factors, including the resolution of replication forks and the disassembly of the replication machinery. Several factors, including chromatin structure, the activity of replication termination factors, and the availability of nucleotides, influence the regulation of replication termination.

Replication stress response: The regulation of DNA replication also involves activating the replication stress response. Replication stress occurs when DNA replication is perturbed due to various factors, including DNA damage, replication fork stalling, or nucleotide depletion. The replication stress response involves the activation of several signalling pathways that coordinate the response to replication stress, including the activation of checkpoint pathways, DNA repair pathways, and changes in gene expression.

Replication fidelity: DNA replication's maintenance of dedication, or precision, is also a crucial aspect of its regulation. The accuracy of DNA replication is of utmost importance in preserving the genetic information within the genome and inhibiting the buildup of mutations that can potentially result in diseases such as cancer. Various mechanisms, such as the proofreading function of DNA polymerases, the involvement of mismatch repair enzymes, and the engagement of DNA damage response pathways, facilitate the maintenance of DNA replication accuracy.

Replication licensing: The topic of replication licensing is being discussed. The regulation of DNA replication encompasses the management of replication licensing, a fundamental process that authorizes replication origins to initiate replication. The method of replication licensing is designed to ensure that each source is replicated only once during a cell cycle. The regulation of replication licensing is crucial in safeguarding against erroneous DNA replication, thereby mitigating the risk of genomic instability and the onset of diseases such as cancer.

Replication-coupled chromatin assembly: The regulation of DNA replication involves the coupling of DNA replication with chromatin assembly. Packaging DNA into chromatin is critical for maintaining the integrity of the genome and regulating gene expression. The collar of DNA replication with chromatin assembly ensures that newly synthesized DNA is packaged adequately into chromatin, which is essential for maintaining the integrity of the genome.

GLOSSARY

DNA polymerase: Enzyme that catalyses the formation of DNA strands by adding nucleotides to a growing chain.

Replisome: A large protein complex responsible for DNA replication.

Helicase: Enzyme that unwinds the DNA double helix by breaking hydrogen bonds between the complementary base pairs.

Complementary strands: Two strands of DNA complementary to each other in base pairing, *i.e.*, A pairs with T and C with G.

Cell division: The process by which a cell divides into two or more daughter cells, each containing a complete copy of the genetic material.

REVIEW QUESTIONS

1. What is the difference between leading and lagging strands in DNA replication?
2. What are the consequences of errors in DNA replication?
3. What is the significance of the semiconservative model of DNA replication?
4. How does DNA replication contribute to genetic diversity?

REFERENCES

[1] Bates D. The bacterial replisome: Back on track? Mol Microbiol 2008; 69(6): 1341-8. [http://dx.doi.org/10.1111/j.1365-2958.2008.06378.x] [PMID: 18673457]

[2] Duggin IG, Wake RG, Bell SD, Hill TM. The replication fork trap and termination of chromosome replication. Mol Microbiol 2008; 70(6): 1323-33. [http://dx.doi.org/10.1111/j.1365-2958.2008.06500.x] [PMID: 19019156]

[3] Duncker BP, Chesnokov IN, McConkey BJ. The origin recognition complex protein family. Genome Biol 2009; 10(3): 214. [http://dx.doi.org/10.1186/gb-2009-10-3-214] [PMID: 19344485]

[4] Groth A. Replicating chromatin: A tale of histones. Biochem Cell Biol 2009; 87(1): 51-63. [http://dx.doi.org/10.1139/O08-102] [PMID: 19234523]

[5] Klinge S, Núñez-Ramírez R, Llorca O, Pellegrini L. 3D architecture of DNA Pol α reveals the functional core of multi-subunit replicative polymerases. EMBO J 2009; 28(13): 1978-87. [http://dx.doi.org/10.1038/emboj.2009.150] [PMID: 19494830]

[6] Lutzmann M, Méchali M. How to load a replicative helicase onto chromatin: A more and more complex matter during evolution. Cell Cycle 2009; 8(9): 1309-13. [http://dx.doi.org/10.4161/cc.8.9.8216] [PMID: 19342892]

[7] Reyes-Lamothe R, Sherratt DJ, Leake MC. Stoichiometry and architecture of active DNA replication machinery in Escherichia coli. Science 2010; 328(5977): 498-501. [http://dx.doi.org/10.1126/science.1185757] [PMID: 20413500]

[8] Bhowmick R, Hickson ID, Liu Y. Completing genome replication outside of S phase. Mol Cell 2023; 83(20): 3596-607. [http://dx.doi.org/10.1016/j.molcel.2023.08.023] [PMID: 37716351]

CHAPTER 4

Mutation and DNA Repairs

Shumaila Yousaf[1,*] and **Saba Nasir**[1]

[1] *Nuclear Institute for Agriculture and Biology, Faisalabad, Pakistan*

Abstract: Mutations are changes in the DNA sequence that can occur spontaneously or due to exposure to mutagenic agents such as chemicals, radiation, or viruses. These changes can have a wide range of effects on the organism, from no effect at all to causing genetic disorders or cancer. DNA repair mechanisms exist to correct these mutations, ensuring the integrity of the genetic material. There are several types of DNA repair mechanisms, including base excision repair, nucleotide excision repair, and mismatch repair, each designed to correct different types of DNA damage. The repair mechanisms are highly regulated and involve a complex network of proteins that detect, remove, and replace damaged DNA. Defects in DNA repair mechanisms can lead to an accumulation of mutations, increasing the risk of cancer and other diseases. For example, individuals with inherited mutations in DNA repair genes have a higher risk of developing certain types of cancer, such as breast and ovarian cancer.

Keywords: Chromosomal aberration, *E. coli*, Gene pool, Genetic material, Genetic variability, Nucleotide.

INTRODUCTION

The molecules in an organism's DNA are not static. Often, the nitrogenous bases of their nucleotides are exposed to agents, natural or artificial, that cause modification in their structure or chemical position. Every organism that exhibits a form different from its ancestor's results from a mutant mutation. In a broad sense, mutation refers to any sudden and hereditary modification in an organism's gene pool, not explicable by recombining the pre-existing genetic variability. These changes include changes in the number of chromosomes (euploidy and aneuploidy), the chromosomal structure (chromosomal aberrations), and changes in individual genes. Within the confines of this particular chapter, the term "mutation" shall be employed in a more limited manner, specifically denoting alterations that are identified solely at the individual gene level. The preservation of genetic material across successive generations relies on the imperative of minimizing mutation rates. High mutation rates in the germline would lead to the

* **Corresponding author Shumaila Yousaf:** Nuclear Institute for Agriculture and Biology, Faisalabad, Pakistan; E-mail: shumailayousuf_alvi@yahoo.com

Sanaullah Sajid, Sajjad ur Rahman, Shahid Mahmood, Shayan Bashir & Mudasser Habib (Eds.)

extinction of the species, while increased mutation rates in somatic lineages would result in the individual's death. In this way, cells depend on the correct functioning of thousands of genes, each of which can be damaged by a mutation in any of the sites of the coding sequence of the protein or functional RNA or regulatory regions that promote their expression. In general, mutation-carrying organisms in a particular gene have problems in their survival and are eliminated by natural selection. However, not every mutation results in a deleterious consequence for its carrier. All genetic variety comes from mutations, which fuel evolution. This variation is rearranged into new combinations through recombination, and natural or artificial selection protects the combinations best suited to the current environment. Mutation allows organisms to evolve and adapt to their environment [1].

NATURE OF MUTATIONS

All living things mutate due to normal cellular functions or random environmental interactions. These mutations are called spontaneous, and their frequency of occurrence is characteristic for each species, constituting the so-called basal level (background). The occurrence of mutations can be increased by treatment with specific compounds. Those compounds are called mutagens, and the modifications they cause induced mutations. Many mutagens act directly on DNA due to their ability to work as a nitrogenous base or to be incorporated into the polynucleotide chain. The effect of a mutagen is measured by the degree to which it increases the mutation frequency above the basal level. Any nitrogenous base of DNA can be mutated. A point mutation involves modification in a single base pair (substitution) or a few pairs of bases (addition/insertion or deletion). Point mutation may result from a malfunction of the cellular system that replicates or repairs DNA by inserting an incorrect base in the polynucleotide chain, which is being synthesized, or from a chemical interference on one of the bases of DNA [2].

CLASSIFICATION OF MUTATIONS

Mutations must cause some detectable modification for their presence to be recognized. Changes can be minimal, identified only by special genetic and biochemical techniques; they can affect the cell's morphology or be lethal. The extent to which a change alters an organism's characteristics depends on where this mutation occurred and the degree to which it affected the functioning of the gene product. Geneticists classify mutations in various ways because of the wide variety of types and effects. These classifications are not mutually exclusive. The same mutation can be classified in more ways than one.

Classification by Location

Mutations can be classified according to the type of cell or the chromosomal location in which they occur; they can arise in any cell cycle stage. Somatic mutations occur in any cell of the body except the germ cells. They will be perpetuated only in the cells descended from the original cell where the mutation occurred and may not affect the entire organism. Germline mutations occur in the gametes and will be transmitted to the offspring. Autosomal mutations are those that occur in genes located on autosomal chromosomes, and X-linked mutations are those that affect the genes located on the X chromosome. If a somatic cell of a diploid organism experiences an autosomal recessive mutation, the probability of manifesting an observable phenotype is low. The phenotypic manifestation of most of these mutations is likely to be suppressed by the wild-type allele within the cell. The phenotypic consequences of somatic mutations are contingent upon their dominance or, in males, their linkage to the X chromosome. Similarly, somatic or X-linked mutations' impact becomes more evident when they manifest early in the organism's developmental process, during which a limited number of undifferentiated cells undergo proliferation to generate diverse tissues and organs. Germline mutations can be phenotypically manifested in all cells of an offspring. A phonetically detectable autosomal dominant mutation will be displayed in the first generation due to inheritance. Mutations occurring in a gamete of a homogametic female, namely those associated with X-linked recessive traits, have the potential to manifest in the male progeny who inherits the afflicted X chromosome in a hemizygous state. The presence of heterozygosity can lead to the transmission of an autosomal recessive mutation in gametes, regardless of gender, even if it entails a deadly allele. This phenomenon may remain undetected through multiple generations until the mutant allele has successfully disseminated within the population. The manifestation of the novel allele will only occur when chance crossings result in pairing two identical copies of the allele in the homozygous state.

Classification by Type of Molecular Alterations

Mutations are often classified in terms of the nucleotide changes that constitute the mutation. As already mentioned, the substitution of a base pair, on the other hand, in a DNA molecule is called a point mutation or base substitution. The exchange of a naked nucleotide within the coding region of a gene can result in the creation of a new codon, which will correspond to a different amino acid in the protein product; this mutation is called a mutation with a missense mutation. One second possible result is that the original codon is replaced by a stop codon, resulting in premature termination of the translated protein called Nonsense Mutation. If the point mutation changes the codon, but this change does not result in the exchange of the corresponding amino acid in the protein (due to degeneration of the genetic code), this is called a silent mutation. Two other terms are also frequently used to describe a base substitution. A replacement of a purine by another purine or a pyrimidine by another pyrimidine is called a transition. The replacement of a purine by a pyrimidine, or *vice versa*, is named transversion. The third type of point mutation involves adding or deleting one or more base pairs. It is collectively called frameshift mutation because it changes the reading phase of all triplet pairs of bases in the gene after the mutated site. A mutation that modifies the reading frame will occur when any of the bases is added or deleted, except for multiples of three, which would restore the original reading frame. One of the many codons changed is a UAA, UAG, or UGA, which are stop codons. When one of these codons is found during translation, the polypeptide synthesis is disrupted. The result of a mutation, which modifies the reading phase, can have serious consequences, primarily if it occurs in the initial portions of the coding sequence.

Classification by Phenotypic Effects

Mutations can have varying effects on phenotype depending on the nature and location of the mutation. When a gene's product undergoes a loss-of-function mutation, its function is diminished or lost altogether. Loss of function can result from any mutation, from a single base-pair change to an entire gene being deleted. Null mutations are those that eliminate a gene's ability to perform its normal function entirely. The presence of a mutated protein product that interacts with or inhibits the action of the normal gene product is one possible outcome of a dominantly acting failure of function mutation. Gain-of-function mutations, on the other hand, cause the encoded protein to perform an additional or different role. This may be due to a change in the protein's amino acid sequence, resulting in recent activity, or the result of a mutation in the regulator region of the gene by increasing its level of expression or synthesis of product at an inappropriate location. However, as eukaryotic genomes consist mainly of non-coding areas, most mutations are likely to occur in large portions of the genome devoid of genes. These mutations are considered neutral mutations if they do not affect gene products or the expression of genes. The most easily observed mutations affect a morphological feature called visible mutations and are known for their ability to alter a specific phenotype.

Some mutations exhibit biochemical or biological effects. In bacteria and fungi, a typical mutation results in losing the ability to synthesize an amino acid or a vitamin. In humans, sickle cell disease and haemophilia result from biochemical mutations. Although these mutations do not always affect morphological characteristics, they can affect the well-being and survival of the affected individual. There is also a category of modification that concerns an organism's behaviour patterns, for example, the mating behaviour of an animal. The primary effect of these behavioural changes is difficult to analyze. Other types of mutations include mutations regulators, which affect the regulation of the expression of a gene by activating or repressing its expression improperly, and lethal mutations, which can disrupt a process essential to the organism's survival. Most of the mutations used in the genetic analysis of biological mutations are conditional lethal mutations within the category of lethal mutations. These mutations are deadly in a given environment, so-called restrictive but non-lethal conditions in another setting, with permissive conditions. The three most common types of conditionally lethal mutants are: (1) auxotrophic mutants, which cannot have the ability to synthesize a necessary metabolite (such as an amino acid, purine, pyrimidine, *etc.*) but has the ability to recover this synthesis from the wild-type members of their species. These mutants can only reproduce in the presence of the metabolite (permissive condition) and die off in its absence (restrictive disease); (2) temperature-sensitive mutants, which can only live in one temperature range. Temperature-sensitive mutants that are temperature-sensitive have a gene product that is usually stable at room temperature but becomes inactive at higher temperatures; (3) suppressor-sensitive mutants are viable only in the presence of a second genetic factor called a suppressor but die off otherwise. The suppressor gene can restore a changed gene product induced by a suppressor-sensitive mutation. The gene product induced by suppressor-sensitive mutation can be fixed by the suppressor gene, or the phenotype it causes can be compensated for.

Spontaneous Mutations

There appears to be no identifiable reason for spontaneous variations in gene nucleotide sequences. These mutations are not linked to a single causative agent, but rather are thought to result from unidentified chemical or biological processes within the organism that cause structural changes to nitrogenous bases.

ERRORS DURING DNA REPLICATION

The replication process of DNA could be better; sometimes, DNA polymerase inserts incorrect nucleotides during the synthesis of a DNA strand. Although DNA polymerases can correct these replication errors, using their ability to inherent error correction in the 3'→5' direction, erroneously incorporated nucleotides may persist after replication. If DNA repair systems do not catch and fix these mistakes, they can lead to mutations. Point mutations are the result of faulty replication caused by mismatched chromosomes. Because DNA bases can exist in multiple configurations (a property known as tautomerization), mismatches are more likely to arise during replication [3].

Replication Slips

DNA replication can also lead to replication slips, which are tiny insertions or deletions of nucleotides. The DNA polymerase can slip or stop at a spot, or a DNA template strand can create a loop and be displaced, both of which lead to these modifications. DNA polymerase can bypass the nucleotides that make up a strap if it exists in the template strand, leading to a minor deletion in the newly synthesized strand. Repeatedly introducing nucleotides not found in the template strand causes DNA polymerase to insert one or more nucleotides, leaving a gap in the freshly synthesised strand. Mutations resulting from insertions and deletions can include those that shift the reading frame or those that add or remove codons or amino acids. However, replication errors tend to cluster in DNA with repeating sequences. DNA mutations tend to cluster around repetitive sequences, making them a potential cause of genetic disorders. Current techniques for forensic DNA analysis are predicated on the hypermutability of repetitive sequences in non-coding areas of the genome.

Tautomeric Modifications

Watson and Crick suggested a mechanism for the occurrence of spontaneous mutations in 1953 in their work on the DNA double helix. They noticed that some hydrogen atoms could move from one position to another in purines. Pyrimidines, for instance, undergo tautomeric changes in which an amino group is attached to a nitrogen ring. Despite their rarity, tautomeric alterations may play an important role in DNA metabolism by disrupting the typical pairing of adenine and thymine or guanine and cytosine. However, the more stable ketone forms of thymine and guanine and the amino forms of adenine and cytosine can undergo tautomeric modifications, respectively. The phenolic and aminic conditions are less durable.

Hydrolytic Damage: Deamination

Spontaneous mutations in DNA arise from errors during replication and injuries caused by environmental mutagens. Even though the double helix structure of DNA itself depends on an aqueous environment, DNA can be damaged by the action of water. The most frequent type and essential feature of hydrolytic damage is the deamination of the cytosine base. Under normal physiological conditions, cytosine undergoes spontaneous deamination, losing its amino group and forming uracil, which is not a base nature of DNA. Uracil pairs with adenine, introducing it into the opposite strand after replication instead of guanine, which should have been paired with the original cytosine. Deamination can occur naturally in adenine and guanine as well. Adenine undergoes deamination to become hypoxanthine, which pairs with cytosine *via* hydrogen bonds instead of Thymine; guanine undergoes deamination to get xanthine, which still pairs with cytosine but *via* just two hydrogen bonds.

The damage generated from forming a base naturally occurring by deamination is illustrated through the problem caused by 5-methylcytosine. Vertebrate DNA often contains 5-methylcytosine instead of cytosine due to the action of methyltransferases. The modified base plays a vital role in regulating the transcription of many genes. Sites containing 5-methyl-cytosine are hot sites for spontaneous point changes. Due to the existence of these sites, 5-methyl-cytosine suffers spontaneous deamination at a substantial frequency; substituting the ketone group converts 5-methyl-cytosine to thiamine. Since this base is one of the usual constituents of DNA, the existing repair system to remove unusual bases in the DNA does not detect the change, resulting in a transition-type mutation. This conversion creates an anomalous pair GT, whose separation, in replication subsequently, produces a wild-type GC pair and a mutant AT pair.

Oxidative Damage

The byproducts of healthy biological functions can potentially wreak havoc on DNA. Electron-donating oxidants, such as reactive oxygen species, are produced as a byproduct of aerobic respiration. Examples of reactive oxygen species (ROS) that pose a continual risk to DNA include the superoxide radical (O^{2-}), hydroxyl (OH) and hydrogen peroxide (H_2O_2). These reactive oxidants, which can also be made by high-energy radiation, can change the bases in more than 100 different ways, which can lead to mistakes when DNA is copied.

Frequency of Mutations

Many generalizations can be made regarding the rates of spontaneous mutations in organisms. The rate of random mutation is relatively low for all organisms. Still, it can vary considerably in different organisms, with values ranging from 10-4 per base pair per generation for certain RNA viruses, from 10-6 to 10-8 for most bacteria, and up to 10-9 for human cells. Even within a species, the spontaneous mutation rate can vary from gene to gene. In general, the frequency of mutations increases with age, an increase in population size and a decrease in the information content of the genome, which results in a very stable mutational rate, approximately 1/300 non-neutral mutations per genome per generation.

However, varying mutation rates throughout the genome complicate this issue. Initial studies on the LAC repressor of *E. coli* revealed a significant variation in the mutation rate across the entire gene length. In contrast, recent genomic studies on silent mutations in humans have shown the presence of hot spots and cold spots that cover hundreds of kilobases. The hot site designation is assigned to sites with several modifications greater than expected for a random distribution, for example, from 10 to 100 times the same mutation in the same site. In opposition, cold sites are places in the genome where the frequency of changes is lower than expected for a random occurrence. In some small genomes of bacteria, some mutational hot sites have been linked to specific DNA sequences, such as inverted or tandem repeats. The reasons for the variation in the frequencies of mutations in genomes are unclear; however, the variation between organisms can reflect the relative efficiencies of its DNA proofreading and repair mechanisms.

Induced Mutations

Every single cell in the body is subjected to a barrage of mutagenic chemicals that can alter DNA. Some of these agents, such as the poisons produced by fungi, cosmic rays, and UV radiation, are naturally occurring components of the environment. Other unnatural modern contributions include cigarette smoke, chemicals used in industrial processes, and medical X-rays.

Base Analogues

Base analogues are mutagenic substances that have a structure similar enough to regular bases to be metabolized and incorporated into DNA during replication. However, they are different enough to increase the mismatch frequency and mutation. The two analogous bases most commonly used are 5-bromouracil and 2-aminopurine. 5-bromouracil (BrdU) is an analogue of thymine that generally pairs with adenine. However, the proportion of the enolic tautomeric form of BrdU is more significant than that of thymine, probably because the Atom of bromine is more electronegative than the methyl group on carbon 5. The enolic form of BrdU pairs with guanine, causing the AT↔GC transition, and 2-aminopurine mostly pairs with thymine. Unlike adenine, the rare tautomer of 2-aminopurine can form a single hydrogen bridge with the cytosine, and thus 2-aminopurine can produce the AT↔GC transition.

Nitrous Acid, Acridine Dyes, and Alkylating Agents

Some mutagens act through chemical modifications of DNA bases. For example, nitrous acid reacts with bases that have amino acid groups. As seen earlier, adenine is oxidatively diamine to hypoxanthine, cytosine for uracil, and guanine for xanthine. Due to the pairing changes caused, the nitrous acid causes AT↔GC transitions. Hydroxylamine (NH_2OH) is a specific mutagen, reacting almost exclusively with cytosine to generate a derivative that pairs with adenine rather than guanine. Hydroxylamine produces a unidirectional transition of GC to AT (Fig. **1**).

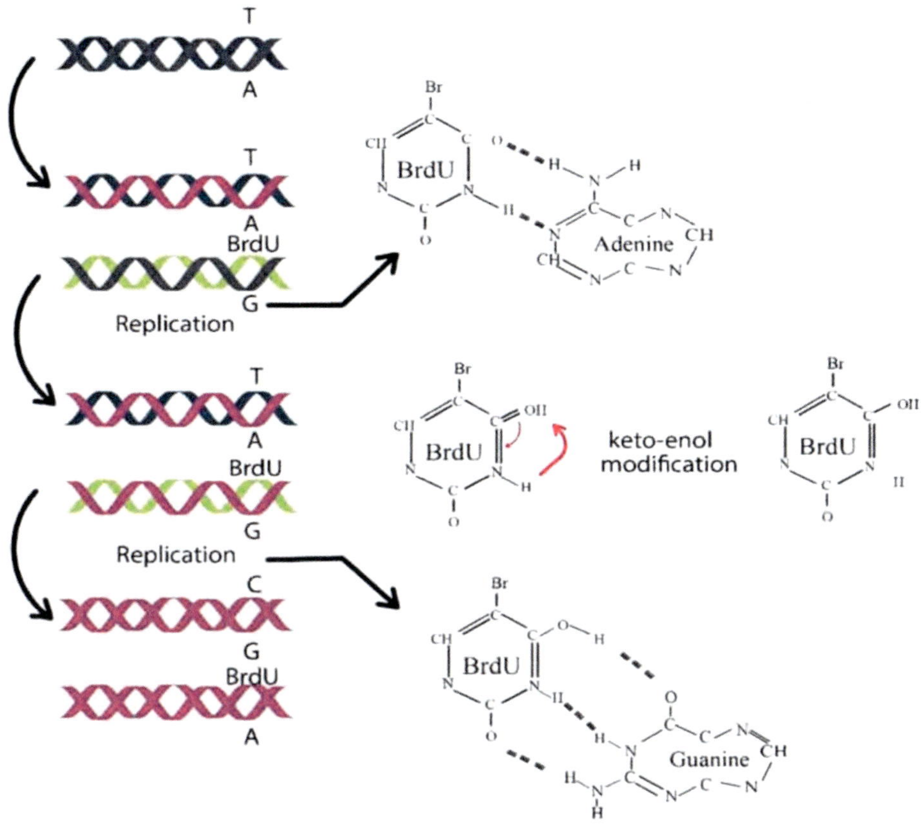

Fig. (1). The nitrogenous base 5-bromouracil (BrdU), the thymine analogue, occasionally pairs with guanine instead of adenine. The Atom of bromine (Br) increases the proportion of rare tautomer formed by displacement of a proton from nitrogen 3 to an Atom of oxygen from carbon 4, causing the AT↔GC transition.

Molecules of aromatic dyes, such as acridines, produce a different type of mutation. These compounds intercalate in the DNA; that is, they are introduced between adjacent base pairs of the double helix and, as a result, cause the insertion or deletion of one or more base pairs. The effect of these mutations is the alteration of the reading frame in translation unless a number multiple of three is entered or deleted. The analysis of these mutants revealed that the genetic code was deciphered in triplets of nucleotides. Another damage from alkylation is the transfer of methyl or ethyl groups to the DNA chain's reactive sites of bases and phosphates. One of the most vulnerable alkylation sites is the guanine base, subject to oxygen methylation attached to its carbon 6. The product of this methylation, an O_6-methylguanine, often pairs incorrectly with thymine, exchanging GC for AT when DNA is replicated.

Radiation-Induced Mutations

The portion of the electromagnetic spectrum containing smaller and higher wavelengths than visible light can be subdivided into ionizing radiation (X-rays, gamma rays, and cosmic rays) and non-ionizing radiation (ultraviolet light). Ionizing radiation, such as X-rays (0.1-1nm), has much energy and is helpful for medical diagnosis, as it can penetrate the tissues. During penetration, these high-energy rays hit the atoms of matter and knock out electrons, leaving behind positively charged free radicals (ions). The subsequent collisions of these ions with other molecules cause the release of more electrons. When high-energy rays pass through biological substance or tissue, they leave behind a "trail" of ions in their wake. Its ionization process is induced by machines that produce X-rays, protons, neutrons, and radiation and are released by the radioactive isotopes of the elements (*e.g.*, P32, S35, Co60, Cs137, *etc.*). Ultraviolet (UV) rays, due to their low intensity, are unable to cause ionization in living organisms and are therefore limited to the outermost layers of plant and animal cells. Upon striking an atom, ultraviolet light causes the electrons to jump to higher energy levels, a process known as excitation. The chemical reactivity of a molecule increases when its constituent atoms are in an ionic or excited state, as opposed to their ground state.

MUTATIONS IN MAN

Sickle Cell Disease

A well-known and well-studied example of a mutation in the human species involves the hemoglobin molecule. Haemoglobin is a carrier macromolecule of oxygen present in red blood cells. Haemoglobin is a complex protein consisting of four globin protein chains combined with the ferric heme pigment. Two main types of hemoglobin are present in males at different stages of their development: fetal hemoglobin (HbF) and adult hemoglobin (HbA). The HbA contains two identical chains composed of 141 amino acids each. At HbF, two chains are present that have 146 amino acids. Due to the similarities in amino acid sequences, it is believed that all hemoglobin chains, therefore, the corresponding genes, have evolved from a common ancestor. Many variants of homoglycans have been identified in human populations, the majority of which are HbA. Some variants have serious phenotypic effects for their carriers, while others present minor results. Sickle cell hemoglobin (HbS) is one of the variants of hemoglobin. Homozygotic individuals for the HbS allele (HBS/HBS indicates that HbS is a chain variant) develop severe hemolytic anaemia, called sickle cell disease (sickle cell anaemia) [4].

HBS molecules precipitate in the absence of oxygen, forming crystalloid aggregates that distort the morphology of red blood cells, stretching and getting sickle-shaped. These cells, in the form of sickles, obstruct the small blood vessels and prevent oxygen transport. In most cases, sickle cell disease causes death during childhood. When the amino acid sequences of the HbA and HbS were determined and compared, it was verified that the chain of HbS differed in only an amino acid (the sixth amino acid from the amino end) about the HbA chain. In the HbA chain, this amino acid is glutamic (negatively charged); in the HbS chain, it is valine (uncharged). The HbA and HbS are identical, so modifying a single amino acid can seriously affect phenotype.

SOS System

Since the cell can regulate the expression of genes according to the need of its products, many enzymes involved in DNA repair are induced by their defect in the nucleic acid molecule. The most important group of these enzymes consists of the effects of the SOS genes caused by a defect that is enough to prevent DNA synthesis instead of just changing the base pairing properties. Pyrimidine dimers are an example of this type of damage. When the replication fork encounters a dimer, the synthesis stops and restarts at some distance beyond the dimer, leaving a gap in the DNA on which the recombinant enzyme RecA binds. Before starting the exchange of the damaged DNA strand, the binding of the protein activates single-stranded DNA in this protein enzymatic activity independent of recombination, which is the stimulation of the autoproteolytic activity of the repressor of SOS genes (LexA protein). In this way, the RecA protein promotes DNA repair of both forms by recombination and induction of about 15 SOS genes. All of these genes are expressed at a significant level, even without induction. Still, its expression level is significantly increased in cells where DNA is damaged since the repressor system (LexA) has been removed [4].

The repression of gene expression by the LexA protein is achieved through its binding to a 20-base pair sequence known as the SOS box. This sequence contains a conserved pattern of eight positions shared among the target genes. Like other operators, SOS boxes exhibit overlapping regions with their corresponding promoters. The LexA locus contains two neighboring SOS boxes that undergo self-repression. RecA and LexA are mutually regulated components within the SOS response pathway. Specifically, RecA serves as an activator for the proteolytic cleavage of LexA, resulting in the downregulation of RecA activity. Conversely, the cleaved form of LexA acts as a repressor for RecA expression. The SOS response leads to the upregulation of both the RecA protein and the LexA repressor, and this outcome does not appear contradictory. The upregulation of RecA expression is imperative owing to its pivotal involvement in recombination repair mechanisms. During induction, the RecA protein exhibits a significant increase in its concentration, reaching approximately 50-fold higher than its initial baseline level of roughly 1,200 molecules per cell. A significant elevation in induced cells indicates that the presence of RecA is adequate to guarantee the cleavage of all LexA proteins. This mechanism is expected to hinder the re-establishment of gene repression by LexA. However, the primary significance of this circuit within the cell lies in its capacity to restore normalcy rapidly. Upon removal of the induction signal, the RecA protein exhibits a diminished ability to destabilize the LexA protein. During this period, the expression of the LexA gene is significantly upregulated. In the absence of Activated RecA, the LexA protein accumulates rapidly in its intact state, leading to the inactivation of the SOS genes. This elucidates the reason behind the high reversibility of the SOS response.

GLOSSARY

DNA (Deoxyribonucleic Acid): The genetic material responsible for encoding the hereditary traits of all living species, with the exception of RNA viruses. Typical human cells have 46 chromosomes, each comprising two strands of DNA. These DNA strands contain up to 100,000 nucleotides, with precise sequences that constitute genes. Deoxyribonucleic acid (DNA) consists of two intertwined chains including linked nucleotides.

DNA Deletion: An abnormality of the genetic material in which a piece of a chromosome is missing. One nucleotide (qv) or an entire chromosome can be deleted. Deleterious effects, sickness, and even health benefits are all possible through deletions.

DNA Damage: DNA can be damaged in a number of ways, and if left unrepaired, this can lead to mutations or a halt in DNA replication. Environmental, physical, or chemical agents can all play a role in causing these variations.

DNA Methylation: A reversible biochemical change to DNA that happens in almost all living things, from bacteria to people. Cytosine (C) can have methyl groups added to it or removed from it using enzymes. It's linked to the suppression of genetic material.

DNA Probe: A fragment of single-stranded DNA that recognizes and binds specifically to another fragment of DNA; the fragment is usually marked in some way that allows it to be detected, such as with a radioactive or fluorescent tag. DNA probes can be used to identify specific sequences in a DNA sample or to identify certain genes.

DNA Repair: The mechanisms responsible for repairing deleterious alterations in DNA, which can be caused by chemical mutagens, are referred to as DNA repair processes. Enzymes facilitate the substitution of individual DNA bases or the replacement of a segment within a DNA strand by utilizing its complementary, paired strand as a template. These processes have the potential to be susceptible to errors and potentially result in detrimental alterations.

REVIEW QUESTIONS

1. Which type of mutagen changes specific nucleotides' chemical structure, causing them to mispair?
2. Which type of mutagen is most likely to cause frameshift mutations?

REFERENCES

[1] Klug WS, Cummings MR, Spencer CA, Palladino MA, Killian D. Concepts of genetics. Upper Saddle River, NJ: Prentice Hall 2003.

[2] Lainé JP, Egly JM. When transcription and repair meet: A complex system. Trends Genet 2006; 22(8): 430-6.
[http://dx.doi.org/10.1016/j.tig.2006.06.006] [PMID: 16797777]

[3] Madhusudan S, Hickson ID. DNA repair inhibition: A selective tumour targeting strategy. Trends Mol Med 2005; 11(11): 503-11.
[http://dx.doi.org/10.1016/j.molmed.2005.09.004] [PMID: 16214418]

[4] McVey M, Lee SE. MMEJ repair of double-strand breaks (director's cut): Deleted sequences and alternative endings. Trends Genet 2008; 24(11): 529-38.
[http://dx.doi.org/10.1016/j.tig.2008.08.007] [PMID: 18809224]

CHAPTER 5

Restriction Enzymes

Zain Ul Abedien[1,*]

[1] *Institute of Microbiology, University of Agriculture, Faisalabad, Pakistan*

Abstract: Restriction enzymes are bacterial enzymes that cleave DNA at specific recognition sequences, usually consisting of four to eight base pairs. These enzymes have become invaluable tools in molecular biology, enabling scientists to manipulate and analyze DNA in various ways. Restriction enzymes are used in various applications, including gene cloning, DNA fingerprinting, and genome mapping. By cleaving DNA at specific sites, restriction enzymes can generate DNA fragments with defined ends, which can then be ligated into vectors for cloning or PCR amplification. Using restriction enzymes in conjunction with gel electrophoresis allows for the separation and analysis of DNA fragments based on their size. There are over 3,000 known restriction enzymes, each with its unique recognition sequence. Many of these enzymes have been isolated from bacteria and are named after the bacterial species from which they were derived. Some restriction enzymes have also been engineered to recognize new recognition sequences, expanding their usefulness in molecular biology. The discovery and development of restriction enzymes have revolutionized molecular biology, allowing scientists to manipulate and analyze DNA in previously impossible ways. As our understanding of the molecular mechanisms of these enzymes continues to grow, they will likely play a critical role in genetics and biotechnology.

Keywords: Biotechnology, Bacterial enzymes, DNA cleavage, DNA fingerprinting, Gene cloning, Genome mapping, Molecular biology, Restriction enzymes, Recognition sequences, Vectors.

INTRODUCTION

The discovery, characterization, and isolation of different types of enzymes have facilitated the research of genomes and their expression, their regulation, and their subsequent correlation with the development of various diseases. Today, the application of molecular biology techniques in multiple areas, particularly in the health sciences, has revolutionized analytical methods, significantly impacting basic and clinical research, especially in diagnosing diseases [1].

* **Corresponding author Zain Ul Abedien:** Institute of Microbiology, University of Agriculture, Faisalabad, Pakistan; E-mail: zainuaf82@gmail.com

Sanaullah Sajid, Sajjad ur Rahman, Shahid Mahmood, Shayan Bashir & Mudasser Habib (Eds.)

Restriction enzymes are a primary tool for the execution of various assays useful in clinical research, from the determination of restriction fragment length polymorphisms (RFLP), as well as in the construction of vectors with recombinant DNA and sequence cloning of DNA from humans, viruses, bacteria, and other microorganisms of interest for the generation of knowledge or application in the medical field. Among enzymes that modify nucleic acids include nucleases, *i.e.*, those capable of cleaving the phosphodiester bonds that link the nucleotides to a chain of nucleic acids. Nucleases are called endonucleases if they can cleave the phosphodiester bond at nucleotides within the nucleic acid chain. They are called exonucleases if they cleave the phosphodiester bonds of nucleotides at the ends of chains. Thus, some 3'exonucleases cleave bonds at the 3' end of the chain, and 5' exonucleases cleave phosphodiester bonds at the 5' end of the chain.

Restriction and other enzymes that modify nucleic acids, such as ligase, allowed the development of recombinant DNA technology. Restriction enzymes exhibit endonuclease activity, are of bacterial origin, and cut phosphodiester DNA bonds at a specific target sequence. Using these enzymes as a biotechnological tool arose from the need to cut the long genomic DNA strands to manipulate and analyse them into smaller fragments.

In 1960, Stewart Linny and Werner Arber isolated enzymes in *E. coli* that methylated DNA molecules and cleaved a phosphodiester bond from the unmethylated DNA. Later, in 1970, the first restriction nuclease characterized by Daniel Nathans and Hamilton Smith, isolated from Hemophilus influenzae, called Hind I. At present, it is known that all species of bacteria synthesize one or more types of endonucleases capable of making cuts in specific nucleotide sequences of an exogenous DNA double-strand; generally, this "foreign" DNA is from bacteriophages, viruses with the ability to infect such bacteria.

These endonucleases are called restriction enzymes because they restrict the permanence of an exogenous DNA in the bacterial cell that produces such an enzyme. The restriction enzymes most frequently used in recombinant DNA technology usually recognize a single sequence of 4 to 8 nucleotides, where they perform the cut.

ORIGIN OF RESTRICTION ENZYMES

Restriction endonucleases are part of the bacterial machinery they rely on to defend themselves against viral infections [2]. The DNA of the bacterium that produces the restriction enzyme is unaffected by its enzyme, as bacteria methylate specific sequences to their DNA to differentiate it from viral DNA utilizing a methyltransferase enzyme at particular bases, which usually are those recognized by their restriction enzymes (Table **1**).

Table 1. Characteristics of restriction enzymes.

Enzyme Type	Energy Expenditure	DNA Cleavage Site	Methylase Activity
I	ATP	Random, outside the target sequence.	Yes
II	-	Specifically in the target sequence.	No
III	ATP	Random, 25 to 27 base pairs downstream from the target sequence.	Yes

Nomenclature

By international agreement, the nomenclature of restriction enzymes is assigned according to their bacterial origin and based on the following rules:

1. Three letters in italics correspond to the bacterium's scientific name from which it was attracted, *e.g.*, *Escherichia coli* (Eco), Hemophilus influenzas (Hin), *etc.* The first letter, in capital letters, corresponds to the genus; the other two to the species.
2. The strain or lineage, if any, *e.g.*, EcoRI, isolated from the"RY13" strain of *E. coli.*
3. In Roman numerals, a number is distinguished if more than one endonuclease is isolated from the same species, *e.g.*, EcoRI, EcoRV.
4. All should indicate an "R" for restriction or "M" for methylase, depending on the enzyme's function, but it is usually omitted.

For example:

Hind III:

H = genus Hemophilus.

In = species influenza.

d = strain DSM 11121.

III = third endonuclease isolated from this organism.

EcoRI:

E = genus Escherichia.

co = species coli

R = strain RV 13.

I = the first endonuclease isolated from this strain.

Classification of Restriction Enzymes

According to their function, restriction enzymes have been classified as types I, II, and III [3].

Type I

Enzymes in this group recognize a specific nucleotide sequence and make a random double-stranded cut in any DNA sequence at an approximate distance of 1000 bp from the cutting site. In addition, they have the function of methylating their genome. These enzymes require adenosine triphosphate (ATP) to move along the DNA molecule from the recognition site to the cleavage site.

Type II

They are enzymes that recognize specific sequences and cut the double strand of DNA at the same recognition site. These enzymes only restrict activity methylation and are used in genetic engineering, cutting DNA into specific recognized sequences. They require $Mg2+$ as a cofactor and do not require DNA. The recognition sequences of these enzymes are palindromic, *i.e.*, they are sequences that are read the same on both DNA strands. For example:

5' GGATCC 3'GATCC 3.'

3' CCTAGG 5'CCTAGG 5.'

Type III

These enzymes recognize a specific sequence and cut the Double-stranded DNA outside the target sequence (approximately 25 to 27 base pairs downstream of the recognition site). They have methylase activity and require ATP to move along the DNA molecule like those of type I.

Applications of Restriction Enzymes

Restriction enzymes are a primary tool in cloning and genetic engineering; they are used, among other purposes, to make restriction mapping of plasmids or genomes. A restriction map is defined as the series of fragments generated by cutting with specific enzyme(s), specific in size (and sequence), which allow them to be used to characterize the DNA in question. This tool is essential for cloning vectors. Also, in determination studies of polymorphs, such as the RFLP technique, depending on whether or not the cutting site of a restriction enzyme is present, the allelic variant can be established, as the molecular weight of the DNA fragment(s) can be of the generated DNA fragment(s) will allow sequence characterization. Likewise, constructing recombinant DNA molecules involves the apparent use of restriction to define and delimit sequences to be inserted into the recombinant construct. Similarly, cutting genomic DNA, which usually depends on the organism, on the order of millions of bp, is more manipulative when these enzymes digest it into smaller fragments. This allows proper handling without the risk of degradation and manipulation for constructing genomic libraries, as discussed in the chapter on cloning vectors [4].

Types of Cuts Produced by Restriction Enzymes

The cuts made by the restriction enzymes on the DNA strands can generate two structures or shapes: cohesive or blunt. The cohesive cuts, also called sticky, are generated because the enzyme cuts in each DNA chain between nucleotides located in different positions concerning the axis of symmetry of the target sequence. These cuts generate single-stranded ends in a segment from 3 to 5 bases in length; the lack of a complementary strand in this fraction facilitates its interaction with another single-stranded fragment in another single-stranded fragment with a complementary sequence, as shown in Table **2**. The ends thus generated promote specificity binding between different DNA segments. Still, they are cleaved by the same enzyme, which gives rise to suitable conditions of specific union for recombinant DNA production. The blunt ends are generated because the enzyme cuts on both DNA strands on the same axis to double-strand the ends. The union between DNA fragments with blunt ends is more non-specific, as it does not depend on the complementarity of single-stranded regions. Fig. (**1**) shows the enzymes that generate cohesive or sticky ends (EcoRI, BamHI, HindIII).

EcoR1 BamHI

Fig. (1). Cohesive or sticky cuts.

Table 2. Cohesive cuts.

Enzyme	Bacteria Origin	Recognition Sequence	Cutting Products
*Eco*RI	*Escherichia coli*	5'GAATTC 3'CTTAAG	5'---G AATTC---3' 3'---CTTAA G---5'
*Bam*HI	*Bacillus amyloliquefaciens*	5'GGATCC 3'CCTAGG	5'---G GATCC---3' 3'---CCTAG G---5'
*Hin*dIII	*Hemophilus influenza*	5'AAGCTT 3'TTCGAA	5'---A AGCTT---3' 3'---TTCGA A---5'
*Taq*I	*Thermus aquaticus*	5'TCGA 3'AGCT	5'---T CGA---3' 3'---AGC T---5'
*Not*I	*Nocardia otitidis*	5'GCGGCCGC 3'CGCCGGCG	5'---GC GGCCGC---3' 3'---CGCCGG CG---5'
*Hin*fI	*Hemophilus influenza*	5'GANTC 3'CTNAG	5'---G ANTC---3' 3'---CTNA G---5'
*Sau*3A	*Staphylococcus aureus*	5'GATC 3'CTAG	5'--- GATC---3' 3'---CTAG ---5'
*Pov*II	*Proteus vulgaris*	5'CAGCTG 3'GTCGAC	5'---CAG CTG---3' 3'---GTC GAC---5
*Sma*I	*Serratia marcescens*	5'CCCGGG 3'GGGCCC	5'---CCC GGG---3' 3'---GGG CCC---5'
*Hae*III	*Hemophilus egytius*	5'GGCC 3'CCGG	5'---GG CC---3' 3'---CC GG---5'
*Alu*I	*Arthrobacter luteus*	5'AGCT 3'TCGA	5'---AG CT---3' 3'---TC GA---5'
*Eco*RV	*Escherichia coli*	5'GATATC 3'CTATAG	5'---GAT ATC---3' 3'---CTA TAG---5'
*Kpn*I	*Klebsiella pneumonia*	5'GGTACC 3'CCATGG	5'---GGTAC C---3' 3'---C CATGG---5'
*Pst*I	*Providencia stuartii*	5'CTGCAG 3'GACGTC	5'---CTGCA G---3' 3'---G ACGTC---5
*Sac*I	*Streptomyces achromogenes*	5'GAGCTC 3'CTCGAG	5'---GAGCT C---3' 3'---C TCGAG---5'
*Sal*I	*Streptomyces albue*	5'GTCGAC 3'CAGCTG	5'---G TCGAC---3' 3'---CAGCT G---5'
*Sph*I	*Streptomyces phaeochromogenes*	5'GCATGC 3'CGTACG	5'---G CATGC---3' 3'---CGTAC G---5'
*Xba*I	*Xanthomonas badrii*	5'TCTAGA 3'AGATCT	5'---T CTAGA---3' 3'---AGATC T---5'

Isoschizomers

Isoschizomers are restriction enzymes obtained from different bacterial species that recognize the same DNA target sequence and cut between different nucleotides of the target DNA sequence (*e.g.*, Asp718 and KpnI). They recognize the 5'GGTAC3' sequence, but ASp718 cuts the link between GG and KpnI cuts between AC nucleotides.

Families of Restriction Enzymes

A family of restriction enzymes does not necessarily recognize the same target sequence but produces similar cohesive ends. An example of a family is BamHI and BglII, which recognize the target sequences 5'GGATCC3' and 5'AGATCT3', respectively, but generate identical cohesive ends: 5'GATC3'.

BamH I: 5'---G GATCC---3'

3'---CCTAG G---5'

Bgl II: 5'---A GATCT---3'

3'---TCTAG C---5'

This feature allows two fragments to cut these enzymes to recombine in a specific manner due to the complementarity of their bases in the cohesive ends generated.

FACTORS AFFECTING THE ACTIVITY OF RESTRICTION ENZYMES

Among the factors that most affect the activity of restriction enzymes are the following activities:

- The biological purity of DNA; the contamination of the sample with other DNA impedes the proper functioning of enzymes.
- Contaminants such as detergents and stabilizers (sodium dodecyl sulfate, Ethylene diamine tetra acetic acid), as high salts and detergents inhibit enzymatic activity.
- Modifications in pH and incubation temperature. Slight variations in these parameters significantly inhibit enzyme activity.
- The degree of DNA methylation, as some enzymes are inhibited by methylation at specific sequences.

Storage Conditions and Preservation of Restriction Enzymes

Restriction enzymes, like many other proteins, may decrease their half-life and activity when improperly stored and handled outside their ideal temperature range; they should be kept at -20°C to avoid freezing and consequent loss of activity and are diluted in a specific volume of glycerol. A container is required to use them, ensuring the least possible temperature variation.

Selection of the Appropriate Restriction Enzyme

All organisms have a restriction map indicating the sites' position offcut for many known enzymes. When cutting is intended to be performed on a known or commercial plasmid, for example, the restriction map is reviewed, which provides an alphabetical list of enzymes that are capable of cutting such DNA, the cutting site, the type of cut, as well as the number and size of each fragment generated. In the case of genomic DNA, it is necessary to know the complete sequence to be worked with to know the cutting sites and predict the number and size of the fragments. It is done by entering the sequence into a specialized database for restriction studies and, in this way, a list of the possible enzymes that recognize the target sequence. The enzyme is chosen, considering the experimental characteristics adjusting to the objective pursued.

Efficiency in the Use of Restriction Enzymes

Instruction sheets or inserts provided by the supplier include the characteristics and description of the optimal activity conditions of the commercially available restriction enzymes. According to the technology that needs to be applied, these conditions are adjusted for the required reaction volume. It must be taken into account the optimum temperature of action, commercial presentation of the enzyme (enzyme units per microliter), incubation time, and possible requirements for additives, such as albumin, ATP, *etc*.

Amount of Enzyme Suitable for an Assay

The amount of enzyme used in a reaction will depend on the concentration of DNA in the sample to be digested. The enzyme concentration is described in the instruction sheet of the supplier, such as units per microliter (u/μl). An enzyme unit is defined as the amount of enzyme required to produce the complete digestion of a microgram of substrate DNA in 60 minutes at the enzyme's optimal temperature of action. The incubation time of the reaction can be increased if the enzyme used is close to its expiration date and its activity decreases. However, the time should not be increased too much because although DNA is a resistant molecule, degradation is risky.

GLOSSARY

Restriction Enzyme: Endonucleases that cut double-stranded DNA into specific sequences (recognition sequences), cleaving sequences (recognition sequences), and breaking the phosphodiester bonds.

Cohesive Ends: This is the name given to the end of a DNA fragment in which one chain is longer than the complementary one and favors the linkage with complementary fragments.

Blunt Ends: This is the name given to the end of a DNA fragment in which both strands end on the same complementary nucleotide.

DNA Ligase: An enzyme capable of joining at its ends two DNA fragments, allowing the formation of a usually circular DNA molecule. It works exceptionally well if both fragments' ends are cohesive and complementary.

Isoschizomers: These are pairs of restriction enzymes specific to the same recognition sequence. For example, SphI (CGTAC/G) and BbuI (CGTAC/G) are isoschizomers of each other.

Restriction Map: Location in a DNA sequence from the cleavage sites of various endonucleases.

REVIEW QUESTIONS

1. What are the storage and preservation conditions of restriction enzymes?
2. How do you determine the ideal restriction enzyme to cut a specific DNA sequence?
3. How is the best efficiency of the restriction enzyme activity achieved?
4. How is the amount of restriction enzyme determined for an assay?

REFERENCES

[1] Glick BR, Patten CL. Molecular biotechnology: Principles and applications of recombinant DNA. John Wiley & Sons 2022.

[2] Lewin B. Pearson Prentice Hall. NJ: Upper Saddle River 2004.

[3] Lehninger AL, Nelson DL, Cox MM. Lehninger principles of biochemistry. Macmillan 2005.

[4] Smith CA, Wood EJ. Molecular biology and biotechnology. DF, Mexico: Addison Wesley Longman 1998.

CHAPTER 6

Cloning and Expression Vectors

Zain Ul Abedien[1,*] and **Rai Shafqat Ali**[2]

[1] *Institute of Microbiology, University of Agriculture, Faisalabad, Pakistan*

[2] *Nuclear Institute for Agriculture and Biology, Faisalabad, Pakistan*

Abstract: Cloning and expression vectors are essential tools in molecular biology that enable researchers to manipulate and study genes and proteins. Cloning vectors are DNA molecules that can carry foreign DNA fragments and introduce them into host cells. Expression vectors are specialized cloning vectors designed to drive the expression of foreign genes in a host cell, allowing researchers to produce large quantities of recombinant proteins. The most commonly used cloning vectors are plasmids; circular DNA molecules that can replicate independently of the host chromosome. Plasmids can be engineered to contain various features such as selectable markers, multiple cloning sites, and regulatory sequences. These features make plasmids versatile application tools, including gene cloning, mutagenesis, and genetic engineering. Expression vectors are typically based on plasmids and contain additional elements that enable the efficient expression of the foreign gene. These elements include a strong promoter, a ribosome binding site, and a transcription terminator, which work together to ensure the production of the recombinant protein. The choice of expression vector depends on the desired protein expression level, the host cell type, and downstream applications. The use of cloning and expression vectors has revolutionized the field of molecular biology, enabling the production of recombinant proteins, genetic engineering of organisms, and gene therapy. However, using these tools requires careful consideration of potential risks, such as unintended genetic modifications or the spread of genetically modified organisms.

Keywords: Complementary DNA, Expression vector, Molecular cloning, Oligonucleotide sequence, Restriction enzyme.

INTRODUCTION

Until the 1970s, DNA molecules were extremely difficult to analyze since it was too long and consisted of only four monomers. At that time, the nucleotide sequence of DNA could only be studied indirectly through the sequence of amino acids in proteins or through their expression in RNA. Therefore, the discovery of restriction enzymes facilitated the study of DNA since it was possible to section

* **Corresponding author Zain Ul Abedien:** Institute of Microbiology, University of Agriculture, Faisalabad, Pakistan; E-mail: zainuaf82@gmail.com

Sanaullah Sajid, Sajjad ur Rahman, Shahid Mahmood, Shayan Bashir & Mudasser Habib (Eds.)

large DNA molecules and separate them into fragments. Each of these fragments was analyzed separately, and it was possible to multiply them through the polymerase chain reaction technique and even determine the sequence of its nucleotides by sequencing. Restriction enzymes can cut off the recognized sequences with specificity. Identifying, isolating, and cloning DNA fragments from different organisms has made it possible to produce recombinant DNA molecules. The term cloning indicates the act of producing many identical copies of a DNA molecule; currently, cloning is used for the production of recombinant molecules and to determine their functions in the organism [1].

CLONING

DNA cloning, or molecular cloning, introduces a DNA fragment known as insert into a DNA molecule called a vector, which can replicate autonomously and independently of the host cell genome. The result is obtaining millions of copies of a recombinant molecule or molecular clone composed of DNA from the insert and the vector. The insert can be DNA obtained from any organism. It can remain as genomic DNA (gDNA), complementary DNA (cDNA), a product of retro transcription of RNA, a PCR product, or a derived RNA by *in vitro* transcription [2].

Cloning Vectors

A vector is a double-stranded DNA molecule that can harbor an exogenous DNA fragment (of another origin).

Classification of Vectors

According to their use, vectors are classified as cloning vectors and expression vectors.

Cloning Vectors: Their purpose is to store sequences and obtain large quantities of the inserted DNA or recombinant molecule. Cloning vectors are usually plasmids, phages, cosmids, and artificial bacterial or yeast chromosomes.

Expression Vectors: The objective is to produce a transcript (RNA) or transcribe the protein product. Expression vectors can be plasmids or phages.

Elements that Form a Cloning Vector: All cloning vectors have, as part of their structure, the following common essential elements:

a) Origin of replication: it is a sequence in the DNA of the vector that provides a single site of recognition to proteins that identify the origin of the replication site

(ORI). Plasmids are characterized by having a single ORI site in their genome and performing unidirectional replication.

b) Selection marker: a gene that generally confers resistance to an antibiotic or generates a particular phenotype with which the cell can be selected, incorporating the vector. Some vector plasmids have a second selectable marker, the lacZ gene, found in the *E. coli* Lac operon and encodes the enzyme β-galactosidase. Thus, when the vector replicates within the host cell, the β-galactosidase can be expressed and generate a functional enzyme. If the cell containing the vector is put into contact with a substrate of the enzyme, such as X-gal (5-bromo-4-chl-ro-3-indolyl β-D-galactopyranoside), is degraded by the enzyme, which produces a blue color that allows identifying the cells that introduced the vector.

c) Multiple cloning site: it is a DNA fragment containing several unique restriction enzyme recognition sites close to each other, with a wide range of possibilities to insert any piece of DNA.

Plasmids

Plasmids are small circular DNA molecules that replicate and are transmitted independently of the bacterial chromosome. They naturally constitute mobile genetic material of bacteria, transporting genes to other bacteria, such as antibiotic resistance. The plasmids remain episomal in the bacterium, that is, without being incorporated into its genome, and there may be multiple copies of the same plasmid in it. When a bacterium reproduces, these Plasmids are transmitted to progeny cells by equal distribution of the bacterial cytoplasm during cellular division. However, they can also be transmitted through the pili between the bacteria. Plasmid replication and transcription depend on the enzymatic machinery of the host cell. Plasmids have been genetically manipulated in the laboratory to preserve only the desirable characteristics and eliminate unnecessary DNA. The plasmids used as a vector, in general, are formed by 2 to 5 kb of DNA, which facilitate the analysis of the inserts incorporated into it. Hundreds of commercially available plasmids offer various possibilities regarding the restriction enzymes used for the cloning, the length of the product to be inserted, and the selection markers of the transformed clones. The most commonly used plasmids are pBluescript, pUC19, pBR322, and pGLO.

Bacteriophages

Bacteriophages are viruses that naturally infect bacteria and behave as obligate intracellular parasites that multiply using the biosynthetic machinery of bacteria. Like plasmids, bacteriophages replicate autonomously, carry genetic information, and can give bacteria new properties or develop pathological processes. To convert it into a vector, the wild-type bacteriophage (natural virus) is genetically modified, consisting of the addition of restriction sites and the elimination of those genes that are not required for replication, which allows the incorporation of DNA fragments of greater length than those plasmids (up to 23 kb). The lambda bacteriophage is the most widely used phage as a cloning vector. Its genome consists of a linear molecule of approximately 45,000 bp, whose ends are 12 nucleotides containing complementary sequences with each other, called cos sites.

Cosmid

Some genomic DNA analysis studies require cloning large DNA fragments. In these cases, cosmids should be used since they can accept inserts up to 45 kb. Cosmids are hybrid vectors of DNA plasmid (provides resistance to antibiotics) and the bacteriophage cos sites (allowing for packaging the DNA). Cosmids derived from the P1 phage can support 85 kb of exogenous DNA. In addition, they contain the selected genes and an ORI site of the plasmid, so they replicate independently of the bacterial genome.

Artificial Chromosomes

Larger DNA fragments (hundreds of kilobases) can be cloned into bacterial or yeast artificial chromosomes, which replicate as an independent chromosome within bacteria or yeast, respectively. These vectors are particularly useful for chromosome mapping studies because they store large fragments. A bacterial chromosome is a construct derived from the F plasmid (fertility plasmid) capable of regulating the equal distribution of plasmids after bacterial division. Usually, it accepts inserts from 150 to 350 kb, but segments up to 700 kb have been cloned.

Yeast chromosomes are artificial chromosomes that contain telomeres, the origin of replication, a yeast centromere, and a selection marker for its identification in cells with a cloning ability of 100 to 1000kb.

Expression Vectors

Expression vectors are used for two purposes: to generate the RNA product of the transcription or to produce the encoded protein in the gene sequence they carry (recombinant protein). The expression vectors are plasmids or bacteriophages in nature.

Expression Vector Components

In their structure, expression vectors have the essential elements in cloning vectors (origin of replication, selection marker, and site of multiple cloning); they also have one strong promoter in their sequence. They are characterized by constituting a simple and effective induction system, having a low basal expression, and being easily transferable to other strains. It must also include termination sequences transcription and polyadenylation tail. Another important factor in an expression vector is the ribosome binding site, which is the Shine-Dalgarno sequence preceding the AUG start codon of translation in prokaryotic mRNAs. This sequence is complementary to the 3' end of the 16S rRNA, whose hybridization allows the assembly of the translation initiation. In addition, the basal expression of the plasmid must be considered, which is known as the number of copies of the plasmid, a characteristic determined by the strength of the site of origin of replication, which gives rise to a high copy number of plasmids. It should also be considered that the presence of promoters on plasmids with a high copy number originates interference with cell viability. Some examples of expression vectors are pcDNA3, pET161, and pDEST10 (Fig. **1**).

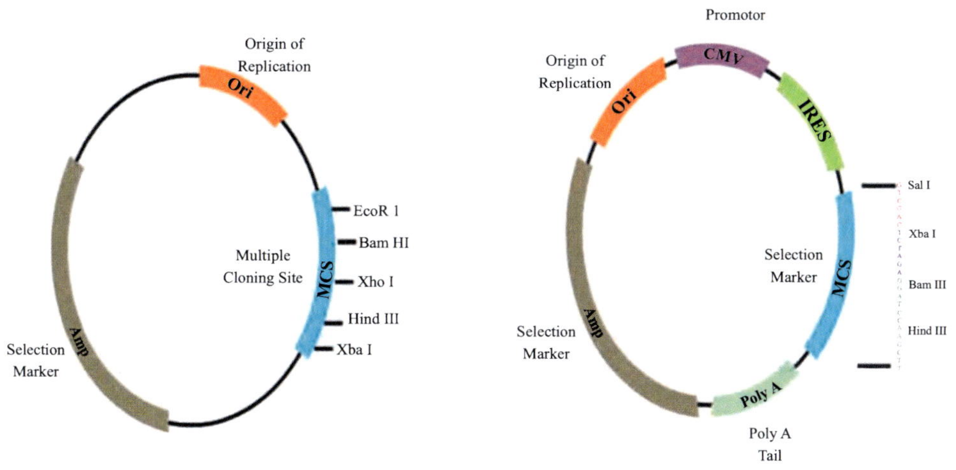

Fig. (1). A cloning vector contains elements such as the origin of replication, a multiple cloning site (MCS), a specific sequence recognized by various restriction enzymes to insert the cDNA of the gene of interest and a selection marker. An expression vector is used to produce a recombinant protein. It contains, in addition to the elements mentioned, a promoter, an IRES site (internal ribosome entry site), and a polyadenylation sequence.

MOLECULAR CLONING

The vector host cell is usually a bacterium (*E. coli*) but can also be eukaryotic. Introducing a vector into a eukaryotic cell is called transfection and transduction if the vector is a virus. The introduction of recombinant DNA is called transformation; genetic material is incorporated extra chromosomally and replicated, transcribed, and translated using the host cell machinery [3]. Cloning any piece of DNA involves 1. Preparing the DNA insert to be cloned. 2.

Preparation of the vector for cloning. 3. Ligation of the insert and the vector. 4. Preparation of competent cells. 5. Cell transformation. 6. Identification of cell colonies containing the recombinant vector.

Preparation of the DNA insert to be cloned: It involves digestion with restriction enzymes to release the fragment of interest from the DNA molecule. The selection of the enzyme(s) that cut the DNA is a critical point that determines the ease and efficiency of the methodology. To facilitate cloning, the use of restriction enzymes whose site cutoffs are present at the multiple cloning sites of the vector, thereby avoiding the need for a possible post-restriction modification of the insert. Likewise, the insert must be cut with the same enzyme(s) or ones that leave the same ends so that the complementarity of the edges generated in the insert coincides with the edges of the vector. It should be taken into account that the incorporation of an insert with blunt ends implies an efficiency of 50% in cloning since half the insert goes in the proper direction (5'-3'). Still, in the other 50% of the cases, the insertion will be in the opposite direction (3'-5'). The insert of the enzymatic digestion is purified and separated from the DNA molecule from which it comes using agarose gel electrophoresis. The band corresponding to the insert (verified by the size it presents) is cleaved from the gel and extracted and purified using conventional techniques.

Preparation of the vector for cloning: The preparation involves the following steps:

a) Cutting with the same restriction enzyme(s) with which the insert was digested and obtaining a linear DNA: The cloning strategy should be designed so that the cut generates ends with complementary sequences to the ends of the insert to be cloned.

b) Dephosphorylation of the vector to prevent self-ligation of the vector: It uses alkaline phosphatase or bacterial phosphatase (BAP), which removes 5' phosphate groups from a DNA strand.

c) Vector purification: It consists of eliminating the agarose particles, the remains of inactivated enzymes, and the salts from the digestion reaction. Generally, it is performed using resin columns with high affinity to DNA in high concentrations of chaotropic salts. This allows the sample elution in small volumes of buffer with a low concentration of salts. There is also the possibility of purifying by the modified phenol: chloroform technique to obtain Plasmid DNA (avoiding genomic DNA extraction); however, it has the drawback that its realization takes a lot of time and that, in general, a lower quality plasmid DNA is obtained. Another ligation strategy consists of the long fragment of the polymerase, which "fills in" the sticky ends generated by any enzyme and makes them blunt.

However, the blunt ends on the vector do not insert directly, so thus the cloning efficiency decreases (Fig. **2**).

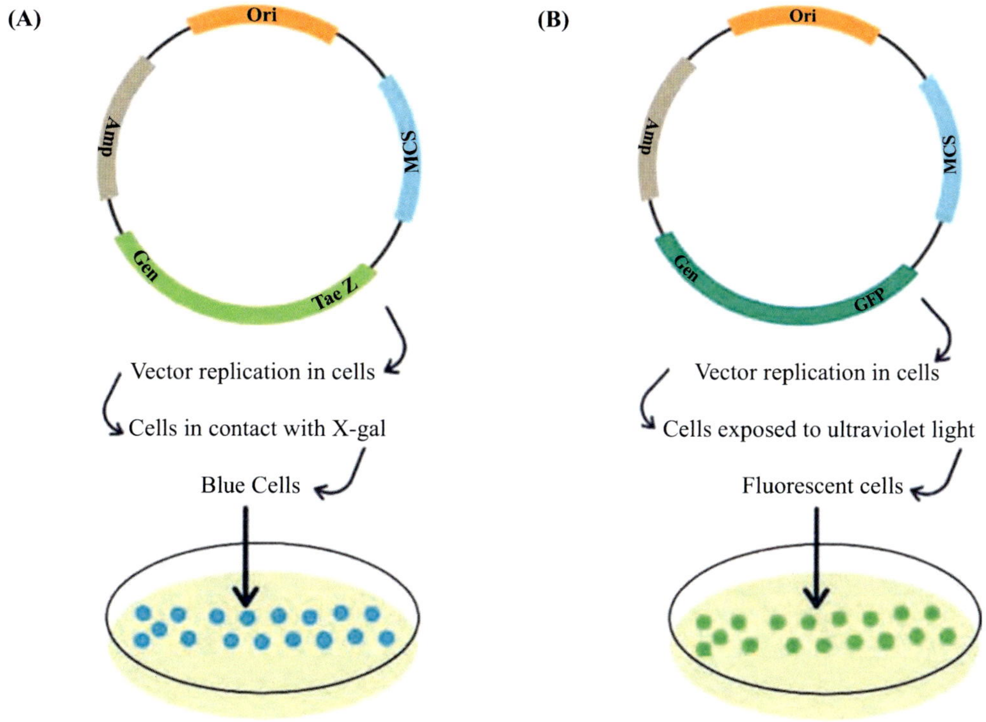

Fig. (2). Selection markers. **A**) The LacZ gene is a marker selection code for the enzyme β-galactosidase. When the cell containing the plasmid comes into contact with the X-gal reagent, the transformed cell colonies are stained blue. **B**) Another selection marker is the GFP gene, which expresses the green fluorescent protein that provides fluorescence to the cell colony when exposed to ultraviolet light.

Ligation of the insert with the vector: The ligation of a foreign DNA segment to a vector involves the formation of phosphodiester bonds between phosphate residues located at the 5' ends of the DNA strand and the 3' hydroxyl residues. This binding is catalyzed *in vitro* by an enzyme called ligase. The most used are *E. coli* ligase and bacteriophage T4 ligase. Different protocols for the ligation reaction depend on the type of ends generated by the restriction enzymes.

Preparation of competent cells: For its replication, vectors need to be inside a cell, which must be prepared to make it permeable to the entry of the recombinant vector, a process called the generation of competent cells, which are usually prokaryotic. The most common bacteria used for this procedure is *E. coli,* which lacks a natural mechanism for incorporating exogenous DNA. Different laboratory methods can induce a state of competence, and the choice of the method will depend on the transformation procedure selected. One of the most used competition methods consists of treating the bacteria with a solution of $CaCl_2$. The exact mechanism by which this compound induces competition is unknown. However, one explanation holds that the bacterial membrane is permeable to chlorine ions but not calcium ions. Hence, molecules of water surround the Ca++ ions to enter the cell, which causes the cell to bulge, which facilitates the formation of pores and the entry of DNA. This procedure is preferred for cells to be transformed using $ClCa_2$ precipitates. Achieving competency status for any of the techniques requires that the cells are in the logarithmic phase of growth (OD between 0.5 and 0.7), corresponding to a cell density of 5×10^7 cells/ml). Another method to confer competition to the bacteria is to subject them to consecutive washings 4-5 times with a 10% glycerol solution, to obtain ion-free cells and protect the cell membrane from damage caused by the electroporation by which they will transform. Regardless of the procedure used to induce the state of competence, cells can be stored indefinitely without losing their competence in a 10% aqueous glycerol solution to prevent them from lysing when thawed.

Transformation into bacteria: Bacterial transformation involves acquiring exogenous DNA, which confers a new phenotype to the host cell. The most common methods are transformation chemistry and transformation by electroporation. It should be noted that the term transformation refers to the introduction of genetic material through vectors to bacteria; The term transduction describes the introduction of exogenous nucleic acid into a cell using a viral vector, and the term transfection implies the introduction of genetic material into cells using chemical or physical agents.

a) Chemical transformation: This technique is based on studies carried out by Mandel and Higa in 1970, who observed that the incorporation of λ phage DNA by bacterial cells was substantially increased in the presence of calcium chloride. Subsequently, this method was applied to introduce plasmid DNA to bacterial cells. In this procedure, the cells and plasmid DNA are incubated in a hypotonic $CaCl_2$ solution, which forms a complex with DNA that facilitates its entry by endocytosis to a competent cell; furthermore, it applies a brief thermal shock at 42ºC for 90 seconds that increases the permeability of the cell membrane. The efficiency of this method is 1×10^4 to 1×10^6 transformed cells/μg of plasmid, which varies depending on the length of the plasmid and the strain of bacteria used. It is a simple method and provides good transformation efficiency for many routine procedures. However, using other divalent cations, such as rubidium and manganese, can increase it.

b) Transformation by electroporation consists of applying brief high-voltage electrical pulses (>1500 V/5 ms) to the receiving cells, forming pores in the plasma membrane in eukaryotic cells and/or the cell wall in bacteria. The membrane components are destabilized, allowing the entry of the vector into the cytoplasm of the cell. On the other hand, this method is the best option for large plasmids of up to 50kb. This is the method of choice when the cells resist traditional transfection and transformation methods. It should be considered that factors such as the length of the vector significantly influence the transformation efficiency, regardless of the technique used for it.

Identification of Colonies Containing the Recombinant Vector: Selectable markers provided are used to identify transformed bacteria by the vectors. The most used are the resistance genes to antibiotics, such as ampicillin, tetracycline, chloramphenicol, and kanamycin. After the transformation, the bacteria are seeded in Petri dishes with agar supplemented with antibiotics, to which the plasmid is resistant. If the transformation has been successful, the bacteria can metabolize the antibiotic and live, indicating that they have incorporated the vector; each colony represents a transformation event only. Other markers' selection is based on the production of fluorescent molecules or enzymes that produce coloration in the transformed cell colony when reacting with a substrate.

APPLICATIONS OF MOLECULAR CLONING

Gene cloning has multiple applications and has been the leading promoter of biotechnology and recombinant DNA technology. This process can be used for the production of eukaryotic proteins in bacteria (recombinant proteins), such as hormones, antibiotics, vaccines, generation of transgenic crops resistant to pests or droughts, production of transgenic animals capable of producing recombinant proteins in their milk, as well as for the establishment of gene libraries, which have allowed the complete sequencing of the human genome and other organisms [4].

Library

A genomic or gene library is a collection of DNA fragments from an organism's genome cloned into vectors to facilitate their study.

cDNA Libraries

cDNA libraries are constructed from mRNA sequences using an *in vitro* reaction synthesizing the first strand of cDNA using a DNA polymerase with reverse transcriptase activity. The second strand of cDNA, complementary to that obtained from reverse transcription, is synthesized in a PCR reaction that uses the single strand of the cDNA as a template. Subsequently, this cDNA double-stranded is cloned into a vector.

gDNA Libraries

To build DNA libraries, isolating the gDNA of the organism of interest is necessary. This gDNA is cut with restriction enzymes to generate fragments with complementary ends to the cloning vectors. The most frequent method of identification of recombinant clones in a library is hybridization using any labeled probe. The use of antibodies against the protein of interest or the selection of clones based on the protein's biological activity. With the advances in recombinant DNA methodology, it is possible to separate specific regions of DNA, sequence them, alter the sequence, and introduce that foreign DNA into the cells of an organism, and even into germ cells, for the production of transgenic organisms.

Recombinant Protein

Molecular cloning has made it possible to determine the function of multiple proteins and generate recombinant proteins that have reduced the costs of hormone production or proteins required in therapeutics, of which previously only small amounts were available due to the scarcity of sources and the difficulty in their isolation or purification. The ease in the production of recombinant proteins has revolutionized the biotechnological area, which is now a dominant market in molecular biology and covers more than 10% of the entire pharmaceutical industry.

Recombinant proteins are produced in bacteria (*E. coli*, *B. subtilis*), yeasts (*Saccharomyces cerevisiae*, *Pichia pastoris*), insect cells (Spodoptera frugiperda), and mammalian cells (hamster ovary cells, baby hamster kidney cell line, HEK293 from human embryonic kidney). To maximize protein expression, strong promoters are used to ensure expression above 10% of the total cellular proteins. However, this imposes a metabolic load on cells by slowing their growth, so specific recombinant proteins can be toxic to the cells that produce them.

The sequence of the gene of interest can also be altered by substituting codons for their rare tRNAs and replacing them with those that code for the same amino acid, which is used more in the host organism. Molecular cloning has increased the initial prospects of recombinant DNA methodology and, thus, in a certain way, it laid the foundations for the cloning of organisms and promoted the development of biotechnology, which has taken a leading role in the health sciences.

GLOSSARY

Cloning Vectors: DNA molecules that can carry foreign DNA fragments and introduce them into host cells.

Plasmids: Circular DNA molecules that can replicate independently of the host chromosome and are commonly used as cloning vectors.

Selectable Markers: Genes that confer resistance to antibiotics or other toxic compounds and are used to select cells that have taken up the cloning vector.

Multiple Cloning Sites: Regions of the cloning vector containing several restriction enzyme recognition sequences, allowing for the insertion of foreign DNA fragments.

Promoter: A DNA sequence that initiates transcription of the downstream gene.

Expression Vectors: Cloning vectors are designed to drive the expression of foreign genes in a host cell.

Recombinant Protein: A protein produced by a host cell genetically modified to express a foreign gene.

REVIEW QUESTIONS

1. What are cloning vectors, and how are they used in molecular biology?
2. What are selectable markers, and why are they essential in using cloning vectors?
3. How is the level of gene expression controlled in an expression vector?
4. What are some of the ethical and safety concerns associated with the use of cloning and expression vectors?

REFERENCES

[1] Berger SL, Kimmel AR. Guide to Molecular Cloning Techniques Methods in Enzymology. London: Academic Press 1987; Vol. 152: pp. 161-522.

[2] Orozco G. Genetics and molecular biomedicine Noriega Limus, Collection of Polytechnic Texts, Biotechnologies Series Sambrook J, Fritsch EF, Maniatis T Molecular cloning: A laboratory manual. 3rd ed. New York: CSHL Press 2001; 1.

[3] Surzycki S. Basic techniques in molecular biology. DNA Cloning general considerations, Chapter 13; Experimental DNA cloning procedures. Berlin, Heidelberg: Springer-Verlag 2000; pp. 298-319. [http://dx.doi.org/10.1007/978-3-642-56968-5]

[4] Sambrook Joseph, Russell David William. The Condensed Protocols from Molecular Cloning. A Laboratory Manual Cold Spring Harbor Laboratory Press. 2006; p. 800.

CHAPTER 7

Recombinant DNA Technology

Farwa Mehmood[1,*]

[1] *Department of Pre-clinical and Clinical Operations, Trials 360 (CRO), Lahore, Pakistan*

Abstract: Recombinant DNA technology, also known as genetic engineering, has revolutionized the field of molecular biology by allowing researchers to manipulate and transfer DNA sequences between different organisms. This technique involves the use of restriction enzymes to cut DNA molecules, which can then be spliced together to create novel sequences. Recombinant DNA technology has numerous applications in medicine, agriculture, and biotechnology. For example, it has been used to produce human insulin, growth hormone, and other therapeutic proteins. In agriculture, it has been used to create crops with improved yield and resistance to pests and diseases. However, the use of recombinant DNA technology also raises ethical and safety concerns, and its regulation is subject to the ongoing debate.

Keywords: Agriculture, Biotechnology, DNA splicing, Genetic engineering, Regulation, Recombinant DNA.

INTRODUCTION

Recombinant DNA techniques comprise a large number and variety of technological developments, which have constantly been advancing over the last 30 years and have given rise to biotechnology, one of the most active and fastest-growing fields of scientific research in the last two decades. These techniques allow researchers to isolate genes from many different organisms, manipulate the purified DNA in the laboratory, and then transfer these genes to other organisms [1].

Gene cloning experiments have survived the hysteria of the 1970s, during which the fears of creating pathogenic organisms causing uncontrollable plagues threatened to suspend this line of experimentation from its origins. These fears have proved to be unfounded since the Asilomar Conference and the development of procedures by the National Institute of Health (NIH) to conduct recombinant

* **Corresponding author Farwa Mehmood:** Department of Pre-clinical and Clinical Operations, Trials 360 (CRO), Lahore, Pakistan; E-mail: farwamehmood4@gmail.com

DNA experiments. On the other hand, as experimental evidence has accumulated the beneficial potential of these new technologies for both basic and applied research, they are now considered one of the most important scientific developments in history. This chapter mainly deals with the basic concepts related to recombinant DNA technology, as well as briefly a description of some related techniques, the most recent advances in this field, and the achievements and future goals of this technology.

Historical background to biotechnology:

- 1909-British physicist Archibald Garrod is the first to propose the relationship between genes and proteins. He suggested that genes may be involved in the formation of the proteins that carry out the reactions of metabolism.
- 1928-Franklin Griffith discovered that increasing temperature could transfer genetic information from dead to living bacteria. This phenomenon was called transformation.
- 1930-Through experiments performed with mutant strains of Neurospora yeasts, George Beadle and Edward Tatum provide evidence supporting Garrod's hypothesis. This evidence gives rise to the hypothesis: “one gene, one protein,” *i.e.*, that each protein in the cell results from gene expression.
- 1944-Oswald Avery, Maclyn McCarty, and Colin MacLeod identify DNA as Griffith's “transforming agent”.
- 1957-During a dysentery epidemic in Japan, biologists discovered that certain strains of bacteria were resistant to antibiotics. It was later found that this resistance was due to plasmids.
- 1961-Sidney Brenner and Francis Crick propose the genetic code, establishing that groups of three nucleotides or codons encode the amino acids of proteins in the DNA.
- 1966- Nirenberg: The genetic code is deciphered by discovering, through biochemical analysis, the amino acid that determines each codon.
- 1970-Hamilton Smith, at the Johns Hopkins School of Medicine, isolates a restriction enzyme for the first time.
- 1972 -Stanley Cohen and Herbert Boyer combine their efforts to generate recombinant DNA technology, giving rise to the biotechnology industry.
- 1976-Herbert Boyer co-founded Genentech, the first company to apply recombinant DNA technology.
- 1978-The first human protein is synthesized using recombinant DNA technology, somatostatin, which regulates growth hormones.

BASICS OF NUCLEIC ACID

The genetic information needed to synthesize the proteins of all cells is contained in DNA (deoxyribonucleic acid). DNA, in the nucleus of eukaryotic, is associated with histone proteins that make up chromatin, a structure with a varying degree of packing that will give rise to chromosomes. In prokaryotes, mitochondria and chloroplasts contain genetic information in a circular DNA molecule. Both cases have two antiparallel strands, which give rise to a double-helix structure. The complementary strands are polymers of nucleotides (nitrogenous base + sugar + phosphate). The genetic information is encoded in the sequence of purines (adenine and guanine) and pyrimidines (cytosine and thymine), which, grouped in threes, make up a codon, or basic unit of genetic information.

Each codon translates into an amino acid and establishes its position in the sequence of a protein. The position occupied by adenine in one strand is paired with thymine in the complementary strand by two hydrogen bonds. Guanine and cytosine pair with each other by three hydrogen bonds. Before cell division, two copies of the DNA must be generated so that each daughter cell retains complete genetic information. This process of producing two identical copies of DNA is called replication or duplication. The information stored in the DNA is transferred to the messenger RNA (mRNA) through transcription. The mRNA is complementary to the DNA strand from which it was copied. The nitrogenous bases that compose it are the same as those of DNA, except thymine, which is replaced by uracil. Therefore, the information in the mRNA will be decoded for the synthesis of proteins through the translation process. This process is carried out in the cell cytoplasm by ribosomes, supra-macromolecular complexes responsible for reading and translating the genetic message into proteins.

The coding region of the DNA contains the codons that will be translated into protein through the mRNA. In prokaryotes, the coding region of DNA is usually uninterrupted and directly produces the mRNA, which practically does not undergo maturation processes and is translated immediately. However, this is not the case in eukaryotes, but the regions of DNA containing the coding sequences (exons) are separated by noncoding sequences (introns). Thus, the initial transcription of the DNA generates a large RNA molecule called heterogeneous nuclear RNA (hnRNA), which will undergo a maturation process inside the nucleus called the adjustment (splicing or editing process), in which the introns will be cut and removed. The exons are spliced together to give rise to the mature mRNA, which will exit the cytoplasm, where it will serve as a template for protein synthesis. The precise function of introns is not yet fully known; nevertheless, it was evident that mRNA production in eukaryotes is much more complex than in prokaryotes due to the negative charges of the phosphate groups;

the DNA fragments move when subjected to the action of an electric field. When this is done through specific gel consistency matrices, the migration speed is inversely proportional to the length of the DNA molecules. This characteristic allows the separation of different DNA fragments by size, which can be visualized after staining the gel matrix with a DNA-specific dye, such as ethidium bromide, for observation under ultraviolet light [2].

RECOMBINANT DNA TECHNOLOGY

These techniques have been aimed at isolating discrete genes from the genomes of living beings, both prokaryotic and eukaryotic, for a study in the laboratory. For handling, a given genome is reduced to a genotype, which consists of a mixture of DNA fragments in which practically the entire starting genome is represented. Genomic libraries are easily and rapidly amplifiable and traceable in search of the genes of interest. The DNA sequence coding for a specific protein represents about one millionth of a eukaryotic genome [3]. To locate it, use the following three basic approaches. First, molecular cloning in a cellular system (genomic libraries) induces the formation of bacterial strains containing a copy of the DNA fragment of interest for subsequent selection. Secondly, hybridization consists of synthesizing a single strand of radioactively labeled DNA or RNA (P32, P33, S32, S33) or chemically (digoxigenin or fluorescein) and its subsequent coupling with the DNA to be identified. Thirdly, the polymerase chain reaction (PCR) consists of obtaining multiple copies of a specific region of DNA by biochemical methods in a test tube.

Several scientific discoveries have been crucial for developing recombinant DNA technology, the original technique introduced by Cohen and Boyer in 1974. The elements involved in molecular cloning are:

1. Plasmids are circular DNA molecules capable of replicating within a bacterium independent of its genome.
2. Insert DNA of interest.
3. Restriction enzymes, endonucleases that cut DNA into specific sequences.
4. DNA ligase is an enzyme capable of end-joining two DNA fragments, allowing circular DNA to form.
5. The transformation process introduces the recombinant DNA manipulated *in vitro* into a host cell for propagation, identification, and isolation.

The restriction endonucleases are obtained from different strains of bacteria, and more than 100 such enzymes have been discovered to date. They possess the unique ability to cut both strands of double-stranded DNA at specific sites. The DNA nucleotide sequence determines the specificity of these enzymes, called the recognition sequence, which identifies around the cleavage site, and usually ranges from four to six nucleotide pairs. A restriction endonuclease makes cutting a DNA molecule at a specific site possible. The recognition sequences are palindromic, *i.e.*, they are read in the same way in the 3′-5′ direction as in the reverse direction on the complementary strand (Table **1**).

Table 1. Restriction site in bacteria.

Bacteria	Enzyme	Restriction site
Escherichia coli	Eco RI	GAATTC CTTAAG
Haemophilus Aegyptius	Hea III	GCCC CGGG

On the other hand, the development of techniques for the synthesis and sequencing of DNA molecules has been fundamental for the advancement of recombinant DNA technology. Recently developed DNA sequencing techniques allow the sequencing of DNA and RNA in a short time. Knowledge of the sequence of nitrogenous bases is important in selecting the appropriate restriction endonucleases. The synthesis of DNA molecules, even those coding for small proteins or peptides, is still arduous. Despite this, these techniques are widely used for synthesizing small DNA sequences (10 to 50 bp) called oligonucleotides, into which the cutting sites (cleavage site) can be incorporated, which is of great importance in the process of gene cloning.

Recombinant DNA techniques make it possible to insert and express small DNA fragments from outside the genome of a living being. In most experiments, the host cells are bacteria. There are two main reasons for using bacteria in these methodologies because of:

1. The relative simplicity of producing mature mRNA and its corresponding protein, compared to a eukaryotic system,
2. The ability of bacteria to multiply exponentially in concise periods makes obtaining large quantities of the DNA or protein of interest possible.

The bacterium most frequently used for this purpose is *Escherichia coli.*

GENOMIC LIBRARIES

Genomic libraries can be of two types, depending on whether the DNA fragments that compose them directly represent the genome of the living being or the mRNA population presents both in tissue and at a particular time of development. In the first case, gene libraries of eukaryotic genomes present some problems, as eukaryotic genes are very large and contain 80 to 90% of the noncoding sequence (introns and regulatory sequences), even if a genomic fragment including a complete gene is cloned into a cosmid, prokaryotes do not possess the enzymes required for the adjustment process, nor do they perform the specific post-translational modifications that are made in both the reticulum system as well as in the Golgi apparatus. Therefore, the expression in a prokaryotic system of a gene cloned in this way is not possible in a prokaryotic system.

On the other hand, in the second case, the genomic library is a collection of complementary DNA (cDNA) (copies of mRNA in the form of DNA) which are propagated in a cloning vector. The cDNA genotypes have three main characteristics. They are sufficiently numerous to adequately represent almost all mRNAs present in cells or tissues from which they were obtained. They contain very few small fragments since, for the most part, they comprise large fragments and may include complete copies of the mRNAs from which they were made. The cDNA libraries may contain the inserts in a specific orientation relative to the transcriptional polarity of the mRNAs or not. Directional gene libraries are usually constructed to favor the expression of the cloned gene by a controllable promoter. Since they can potentially express an antigen, the individual clones can be isolated by differential screening using antibodies that detect the expression of the corresponding protein.

On the other hand, the components of an unaddressed gene library include inserts cloned in either orientation (sense or antisense); only 50% of the clones will be in the direction suitable for expression. Gene libraries can be generated in different ways and with different vectors. The construction of a sound cDNA library starts with obtaining high-quality mRNA. To achieve optimal results, mRNA must be purified on an affinity column using a cellulose oligo(dT) matrix, preferably selecting polyadenylated messenger ribonucleic acid (poly-A+ mRNA).

The amount of mRNA needed to prepare the gene library depends on the efficiency of the individual steps required to convert the mRNA into a form that can be cloned, as well as on the efficiency with which recombinant molecules can be introduced into a vector. The cloning vectors are diverse and must meet certain characteristics: they must possess an origin of replication (which gives them the ability to be replicated independently of the genome of the cell containing them), have a marker that allows identification of the cells that contain them, restriction sites (cutting sequences identified by specific restriction endonucleases) compatible with those in the fragment to be inserted, and to have a mechanism for incorporation into the host cell once constructed.

CLONING

The cloning procedure of a gene or DNA fragment is a technique in which a DNA sequence is introduced into a virus or bacterium, which is then cultured in conditions conducive to their maximum growth and has the following steps:

1. Initial isolation or synthesis of the DNA fragment of interest.
2. The coupling of this sequence to a carrier or vector means a DNA molecule capable of replicating independently of the genome within a cellular system.
3. Incorporating the vector-DNA complex into the host cell is a procedure called transformation.
4. Expression of the DNA in the host cell.

The introduced DNA sequence will then be replicated along with the natural DNA during growth. When the required amount is reached, the DNA sequence can be extracted equivalent to the number of bacteria or viruses produced, achieving amplification of the sequence. The cloning technique relies on a group of restriction enzymes endonucleases obtained from bacteria. The DNA of interest and the plasmid used as a vector are digested with a restriction endonuclease to clone a sequence, generating compatible fragments. In cloning sequences to induce their expression, when the vector with the insert is introduced into the cells where it will be expressed, they will keep the transcript repressed or inactive until a substance that activates the promoter is added to the medium that activates the promoter (Table **2**).

DNA PREPARATION

The DNA of multicellular living beings contains the coding information for an enormous number of proteins, so locating the particular sequence that codes for a specific protein, for example, hemoglobin, is extremely difficult. In addition, DNA includes a large proportion of noncoding sequences (introns) inserted between the coding sequence (exons). If such a DNA fragment were inserted and transcribed directly into a bacterial system, the bacterium would not be able to process the immature precursor mRNA and the corresponding protein would not be synthesized because it does not possess the necessary enzymatic machinery. Therefore, to produce a particular protein in a bacterial system, it is essential to incorporate the definitive sequence of the mRNA in the form of a DNA sequence from which the bacterium could transcribe the appropriate messenger and even translate it into the corresponding protein.

Table 2. Detailed analysis of cloning vectors.

Cloning Vectors
Plasmid Vectors *Advantages* They have an origin of replication that allows multiple copies to be made in the same cell (100 to 200 copies per cell). They have selection markers. Resistance to ampicillin through the production of the beta-lactamase enzyme that inactivates penicillin by proteolysis. Resistance to kanamycin through the production of a phosphotransferase that phosphorylates kanamycin, preventing its binding to the ribosome. Identification of positive clones. The insertion site eliminates resistance to tetracycline by using plasmids with a double resistance marker (pBR322). Transformants with inserts are resistant to ampicillin but sensitive to tetracycline, whereas transformants without inserts are resistant to both antibiotics. The insertion site cleaves the lacZ gene using a chromogenic marker (pUC plasmids). Transformants with inserts are white, whereas transformants without inserts are blue. They include unique restriction sites and are designed with multiple cloning sites (Polylinker), which increases the flexibility in terms of the selection of restriction enzyme(s) used for cloning (pBluescript or pUC18). Designed with a specific promoter at both ends of the insertion or cloning site, which allows transcription of the insert in both directions. Engineered with bacteriophage replication origins to produce single-stranded DNA at both ends of the insertion or cloning site, which can be used for sequencing. *Disadvantages* They allow cloning only of small fragment inserts of 4 to 5 kb on average, with a maximum of about 15 kb. Vector Average insert size Plasmid 4-5 kb Lambda 15-20 kb Cosmid 40-45 kb YAC 400 kb
Lambda Bacteriophage Vectors **Construction:** Using a restriction enzyme, the central portion of the bacteriophage lambda genome is removed (about 20kb), which codes for the capsid proteins. The left and right arms, which carry the genes required for replication and packaging in the phage head, are isolated.
Insertion: The genomic DNA to be cloned is partially digested, generating large fragments, which increases the possibility that they may contain entire intact genes. Then, the phage arms and these fragments will be ligated, forming a long chain linked by coadjacent sites. This concatenates with extracts of the capsid packaging proteins, which self-assemble *in vitro* into the phage heads. Host cells are infected with the bacteriophage thus constructed. Each lysis plate represents a different fragment of the genotype. *Advantages* They allow the cloning of large fragments (average 15 to 20 kb and up to 50 kb). They have a high transformation efficiency.

(Table 2) cont.....

Cosmids Vectors **Construction**: A pair of cos sites is introduced into a plasmid used as a vector and cut with a restriction enzyme at a site adjacent to the cos sites. The genomic DNA is partially digested with a complementary restriction enzyme, and the product of both digests is ligated to construct a large concatenate. The concatenated DNA is then mixed with an extract of phage lambda capsid proteins, which self-assemble by packaging the DNA at the ends of the phage. Cells are then infected; instead of forming plaques, the lambda DNA is recirculated to construct a plasmid. The transformants are selected for the antibiotic resistance conferred by the plasmid. Although the plasmid is very large, the selection pressure keeps it in the bacteria. ***Advantages*** They allow the cloning of large fragments (more than 45 kb), which increases the probability of cloning complete eukaryotic genes. Allow high efficiency of incorporation of large plasmids into bacteria.
Yac Vectors (Yeast Artificial Chromosomes) These are linear artificial chromosomes. They are constructed with large DNA fragments (up to ~1000 kb). They contain yeast ARS sequences for replication. Contain yeast CEN sequences for segregation. Contain yeast telomeres for chromosome stability. For identification, contain yeast markers (one on each arm of the chromosome).

Reverse transcriptase is an enzyme capable of synthesizing a DNA molecule from an RNA template. The cDNA produced in this way is identical to the coding portion of the DNA from which the RNA in question was originally transcribed. Once the cDNA is obtained, the complementary strand (second strand), separating the cDNA from the original mRNA, which is digested, and proceeds to the synthesis of the second strand through a DNA polymerase enzyme, resulting in a double-stranded cDNA representative of the original mRNAs, which can be incorporated into the bacterial system by insertion into a suitable vector.

This system has its limitations, as it is not able to achieve, for example, post-translational modifications, which are indispensable for many proteins to be functional; however, it does, in most cases, yield large quantities of protein, which can be purified for various applications, *e.g.*, the production of antibodies. Another system for producing a given DNA molecule is direct chemical synthesis, which is only possible and affordable for very small molecules. It is essential to know the molecule's sequence to be reproduced by synthesis.

Construction of the Plasmid or Complex DNA Vector-Inserts

Once the DNA to be cloned has been produced, it must be incorporated into a system where it can be expressed. This involves the insertion of the fragment into a vector, for which it is essential to elaborate a restriction map, which allows locating the number and position of the restriction sites of the endonucleases to be used. The vectors most often used are bacterial plasmids, *i.e.*, small circular DNAs that are not part of the bacterial genome and can replicate autonomously from the bacterial genome [4]. Therefore, when the plasmid replicates, the fragment inserted into it will also be copied. Since the bacterial system can replicate, transcribe, and translate the information contained in these plasmids, at least in theory, it is possible that the same can be inserted into the bacterial system to produce the corresponding protein. One or more restriction enzymes can be used to achieve the cloning or insertion of the DNA fragment of interest into the corresponding vector. One way to do this is by first identifying a unique cutting sequence in the vector, which allows the vector to be opened to insert the corresponding fragment using a DNA ligase.

Of course, there must not be a cut site identified by that enzyme within the insert since it would split it, preventing its insertion. Small fragments of DNA containing various cleavage sites for specific restriction endonucleases can synthesize chemically and then ligate to both ends of the DNA to be cloned or even in the vector itself. When a restriction endonuclease acts, two types of ends are produced on the DNA molecule: blunt ends, when the cut is made exactly at the same nucleotide on both strands, and the cohesive ends, when the cohesive ends, when the cut is made at different distances on both strands (one to five nucleotides apart), one strand is, therefore, longer than the other. In both cases, the molecule (insert) that will later be covalently bound must have the same type of cleavage, either with blunt or cohesive ends, to bind correctly. The efficacy of the cloning process will depend on several factors, the most important of which are: the complete action of the restriction endonuclease (fragment digestion), the relative sizes and quantities of insert and vector, the activity of the ligase, as well as the type of ends to be ligated.

Transformation

Once the vector-insert complex has formed, it must be incorporated into a bacterium. This procedure is called transformation. The precise mechanism of incorporating such a complex through the bacterial membrane is not yet fully known. However, it is known to involve a transient change in membrane permeability. It is a slow and inefficient process; once this is done, it must be carefully determined which bacteria have incorporated the plasmid and which have not. The most common method to monitor the transformation process is using plasmids containing a gene that confers specific resistance to certain antibiotics.

When transformed, the bacteria are grown in a selective medium in the presence of the corresponding antibiotic. Only transformed bacteria will be able to grow under these conditions. This procedure is performed in Petri dishes so that each colony obtained originated from a single transformed bacterium. All the bacteria in that colony are derived from it so that each one constitutes a clone since all the bacteria that make it up are identical. This process is called cloning.

As mentioned, not all the plasmids obtained after the insertion and ligation process contain DNA fragments, so it is necessary to select them. On the other hand, in the gene cloning experiments in which we start from mRNA to produce double-stranded cDNA, the population of molecules used is also very heterogeneous; therefore, in these cases, you must make a careful selection of the bacterial clone that contains exactly the insert of interest and no other. Various procedures have been designed to detect the clone of interest and its product, a protein (Fig. **1**).

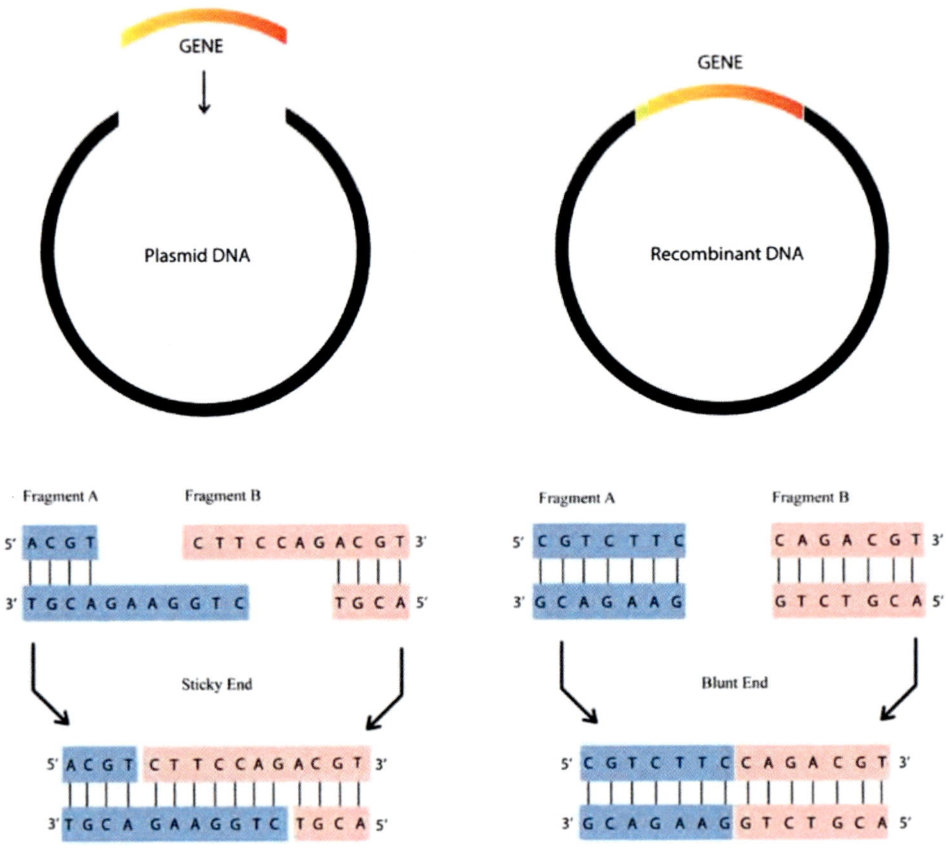

Fig. (1). DNA molecule with two ends.

If the DNA fragment does not code for a protein, the most frequently used method to identify the clone of interest is labeling a complementary fragment to the one of interest. Such labeling can be radioactive, P32, or chemical, digoxigenin. The DNA of the clones will be extracted and hybridized with this probe, allowing the specificity of the corresponding fragment to be identified. On the other hand, if the insert (gene of interest) codes for a protein, the choice of procedure will depend on the properties of the protein. For example, if the protein to be identified catalyzes a specific enzymatic reaction or has some biological activity, we can detect it. Another widely used method is labeled antibodies, either radioactive or fluorescent. These antibodies can be prepared by injecting an animal with the protein against which they are to be obtained. The immune system will recognize this protein as foreign to it and then produce antibodies that recognize it in a specific manner. The antibodies thus produced can be isolated and labeled. Gene cloning experiments are also used to purify specific DNA fragments. If only a small amount of a certain DNA fragment can be obtained by synthesis or isolation, its cloning can help in getting large amounts of that fragment. Once a bacterial clone is obtained containing the fragment, it is possible to obtain practically any number of copies of the fragment from this clone.

Expression of Cloned DNA in a Bacterial System

For the cloned DNA to be expressed in a bacterial system, it is necessary to consider certain additional aspects. The DNA to be expressed must be cloned into a plasmid (vector) recognizable by the enzyme in charge of mRNA production by transcription [5]. Two specific recognition sites are required for DNA expression in a bacterial system:

- The promoter.
- The site where the RNA polymerase in charge of transcription binds.
- The ribosome binding site.

This site must be present in the mRNA to allow its binding to the ribosome and its protein translation. For its part, the efficiency of the expression process will depend on several factors:

- The number of copies of plasmid per cell
- The efficiency of the promoter
- The frequency rate of codon usage
- The stability or half-life of the mRNA
- The level of proteolysis of the host cell

Thus, it is both science and art to successfully construct a plasmid to correctly express the cloned DNA in a bacterial system. In most cloning experiments, knowing the exact sequence of both the vector and the fragment to be cloned and the position and number of cleavage sites for a series of restriction enzymes is essential. This will allow us to manipulate the different DNA fragments. In other words, using DNA synthesis and sequencing systems combined with certain restriction enzymes, it is possible to construct, clone, and express practically any DNA molecule in a bacterial system, or at least in theory.

APPLICATIONS OF RECOMBINANT DNA TECHNOLOGY

From what has been presented so far, it is clear that one of the primary uses of recombinant DNA technology is the large-scale production of clinically essential substances. This is the case of interferon, which has excellent antiviral potential. However, obtaining even minimal quantities of human interferon had been very difficult. It is necessary to get large quantities of any substance tested for therapeutic purposes, to carry out many experiments before its use and for its clinical application.

In this regard, recombinant DNA technologies provide researchers with powerful tools to produce virtually inexhaustible quantities of, in theory, almost any protein with biological activity, from a small peptide to a high-molecular-weight protein, as is the case with such clinically important proteins, in addition to interferons, such as insulin and growth hormone. On the other hand, some genetic conditions involving the production of a specific deficient protein are amenable to treatment by replacement therapy for that enzyme. The limiting factor is the amount of the enzyme that can be obtained and provided. Theoretically, recombinant DNA technology can generate large quantities of enzymes for this and other purposes. Among the cause of many diseases, such as sickle cell anemia, Tay-Sachs disease, and others, is a defect in a single gene. Replacement of the defective gene with a functional version would lead to a complete cure of the condition. Until recently, this treatment was certainly improbable, as it would involve the insertion of DNA into a eukaryotic cell and its subsequent expression.

The DNA of multicellular eukaryotes is much more complex than that of bacteria because it is permanently associated with proteins (histone and non-histone) and their different levels of structural organization and compaction. Nevertheless, recently, Mantei, Boll, and Weissman have cloned the DNA of the β-subunit of hemoglobin. They inserted this DNA into mouse cells, and the DNA functioned correctly and produced functional mRNA. In addition, Cline and co-workers have also succeeded in transforming bone marrow cells with the DNA of the dihydrofolate reductase enzyme, making them resistant to methotrexate. They reintroduced the cells to the bone marrow and observed that the cells not only behaved normally but were able to produce the enzyme, which was functional. In both experiments, DNA coding for a specific protein was inserted into a eukaryotic cell and functioned in a normal manner, *i.e.*, it could produce a translatable mRNA into a functional protein.

As recombinant DNA techniques become more refined and the more we learn about the mechanisms of DNA function in the cells of multicellular living organisms, the more we will be able to perform the complex genetic manipulations currently carried out in bacteria. These studies will gradually lead to a stage where gene replacement therapy will be almost systematic. Recombinant DNA techniques are responsible for discovering spacer sequences (introns) in the DNA of multicellular living organisms. They will undoubtedly play a crucial role in understanding how they function. The potential applications of recombinant DNA technology, both in basic and applied research, seem to be limited regarding the researchers' imagination [6].

GLOSSARY

Cloning: At the molecular level, it consists of inducing, in a cellular system, the formation of a bacterial strain containing a copy of a DNA fragment of interest in the form of a plasmid and its subsequent selection.

Recombinant DNA: DNA molecule synthesized *in vitro*, without the intervention of a cell, by joining DNA fragments of natural or synthetic origin, which can replicate within a cell independently of its genome.

Restriction Enzymes: Endonucleases that cleave double-stranded DNA into specific sequences (recognition sequences), cleaving phosphodiester bonds, *i.e.*, EcoR1 cuts at the GAATTC sequence and BamH1 cuts at the GGATCC sequence. In bacteria, they are enzymes used for protection, mainly for bacteriophages, because they degrade exogenous DNA, unlike their own, modified by methylation in these sequences.

Episome: A fragment of extrachromosomal DNA that is incorporated into a chromosome. The incorporated DNA can come from a virus or a plasmid. Although some authors use it as a synonym for plasmid, others consider it a virus, considering bacteriophage λ (lambda) as an episome.

Cohesive Ends: This is the name given to the end of a DNA fragment in which one strand is longer than the complementary strand, favoring ligation with complementary fragments.

Blunt Ends: This is the end of a DNA fragment in which both strands end with complementary nucleotides.

Insert: DNA of interest will be incorporated into the plasmid used as a vector.

DNA Ligase: An enzyme capable of joining two DNA fragments, allowing the formation of a generally circular DNA molecule. It works particularly well if the ends of both fragments are cohesive and complementary.

Gene Manipulation: The formation of new combinations of hereditary material by *in vitro* insertion of DNA fragments into a virus, plasmid, or any other vector that allows its incorporation into the genome of a host cell or organism in which such a fragment was not naturally occurring before, but usually propagates after that.

Restriction Map: Location, in a sequence of DNA, from the cleavage sites of various endonucleases.

REVIEW QUESTIONS

1. Discuss the contribution of recombinant DNA technology in healthcare.

2. What is the role of recombinant DNA technology in gene therapy?

REFERENCES

[1] Glick BR, Patten CL. Molecular biotechnology: Principles and applications of recombinant DNA. John Wiley & Sons 2022.

[2] Berg P, Mertz JE. Personal reflections on the origins and emergence of recombinant DNA technology. Genetics 2010; 184(1): 9-17. [http://dx.doi.org/10.1534/genetics.109.112144] [PMID: 20061565]

[3] Li MZ, Elledge SJ. Harnessing homologous recombination *in vitro* to generate recombinant DNA *via* SLIC. Nat Methods 2007; 4(3): 251-6. [http://dx.doi.org/10.1038/nmeth1010] [PMID: 17293868]

[4] Vajo Z, Fawcett J, Duckworth WC. Recombinant DNA technology in the treatment of diabetes: insulin analogs. Endocr Rev 2001; 22(5): 706-17. [http://dx.doi.org/10.1210/edrv.22.5.0442] [PMID: 11588149]

[5] Berg P, Singer MF. The recombinant DNA controversy: Twenty years later. Proc Natl Acad Sci USA 1995; 92(20): 9011-3. [http://dx.doi.org/10.1073/pnas.92.20.9011] [PMID: 7568062]

[6] Watson J. Recombinant DNA. New York: Freeman and Co. 1992.

CHAPTER 8

DNA Polymorphisms and Genetic Fingerprint

Itrat Fatima Toor[1,*]

[1] *National Institute of Food Science and Technology, University of Agriculture, Faisalabad, Pakistan*

Abstract: DNA polymorphisms are variations in the genetic sequence that occur within a population. These polymorphisms can be used as genetic markers to identify individuals, determine familial relationships, and study population genetics. Genetic fingerprinting is a widely used method for identifying individuals based on DNA polymorphisms. Genetic fingerprinting involves the analysis of DNA polymorphisms at multiple loci to generate a unique genetic profile for an individual. The most common types of DNA polymorphisms used for genetic fingerprinting are short tandem repeats (STRs) and single nucleotide polymorphisms (SNPs). STRs are short, repeated sequences of DNA that vary in length between individuals. They are highly polymorphic and generate DNA profiles by analyzing the number of repeats at each locus. SNP markers, on the other hand, are single nucleotide variations that occur at specific positions within the genome. They are less polymorphic than STRs, but genetic mapping and association studies are widely used. Genetic fingerprinting has many applications, including forensic science, paternity testing, and conservation biology. However, the use of genetic data also raises ethical concerns regarding privacy and discrimination.

Keywords: Conservation biology, DNA profiles, DNA polymorphisms, Genetic fingerprinting, Paternity testing, Short tandem repeats (STRs), Single nucleotide polymorphisms (SNPs).

INTRODUCTION

The DNA of all known species of organisms has the same chemical structure; however, each organism is entirely different from another; the difference is due to the order of the nitrogenous bases in the DNA molecule. Organisms of the same species share sequences in their DNA molecule, but within the same species, there are variations between individuals. Organisms of the same species will share up to 99% of their sequences, which gives them similar traits. Close relatives will also

* **Corresponding author Itrat Fatima Toor:** National Institute of Food Science and Technology, University of Agriculture, Faisalabad, Pakistan; E-mail: itratfatima71@gmail.com

Sanaullah Sajid, Sajjad ur Rahman, Shahid Mahmood, Shayan Bashir & Mudasser Habib (Eds.)

have more similar sequences, but they will never be identical. This is what is meant by intraspecies and interspecies genetic variability.

In human DNA, the gene sequences are not very variable within the species, but the rest of the sequence is very variable. Due to many millions of base pairs per DNA molecule and a high percentage of it does not code for a protein, each person has a unique DNA sequence that can be identified only by its base pair order, called polymorphism. There are two types of genetic polymorphs: those that show a single nucleotide change by substitution of bases and those involving changes in the size of the sequence; this may be due to insertions or deletions of DNA sequences or to base repetition (or combination of bases) continuously in a segment of the DNA.

This variability is a typical biological process since 30% of the DNA sequence is highly repetitive, which makes it possible to establish specific genetic patterns to identify individuals, *i.e.*, to establish a genetic footprint or DNA fingerprinting. Therefore, scientists have developed strategies to identify individuals by analyzing repetitive patterns of DNA, which allow relating samples from the exact origin or relative people (common ancestors). These sequences were chosen because they are known to vary between unrelated individuals, so the analysis of a group of them to find concordances allows one to infer associations with high probabilities of having the exact origin or being from the same person. A gene is a chromosomal region identical to a DNA segment with a predetermined sequence and contains the instructions needed to produce a particular protein. An organism's genes contain accurate information about every part and process of a person's training and development. However, genes cannot control behavior, so it is important to remember that the environment plays a significant role in how these genes are expressed throughout an organism's life. A gene presents in a double strand, one that provides the father and the other the mother; these copies are called alleles, so there are two alleles for each gene, one of maternal origin and another paternal. These allelic forms of the same gene exist because the DNA sequence is subject to changes, sometimes due to mutations that happen randomly but that give rise to alternate, stable, and inherited forms of a gene with a different function than the original allele or wild. The mutation is the basis of the gradual variation of the genetic structure of populations; that is, the base of evolution. The physical position where a gene is located in the genome is called a locus (plural loci); therefore, the wild-type or modified alleles of the same gene reside in the same locus. When an individual has different allelic forms at the locus is said to be heterozygous since each allele's sequence can generate variants of the same protein. The protein product will be the same regardless of whether the allele is wild-type or modified. On the other hand, an individual is to be homozygous when they display the same allelic form (of either their paternal or maternal origin) at the locus in question [1].

POLYMORPHISMS

Organisms have many variable sequences, and population genetic analyses have made finding two or more alleles for each one possible. The more different the alleles are, the more polymorphic they are. It means it is less likely that two people will have the same polymorphic regions, making it easier to tell them apart. It considers that there is a genetic polymorphism when there are multiple alleles of a gene in a defined population or, more specifically, when the base sequence neogenesis of the DNA molecule of a particular locus is variable among the organisms of a population. The word polymorphism is composed of poly (many), morpho (form), and is (process or state), *i.e.*, “in many ways”. A polymorphism can be observed in a complete individual, in varying forms of proteins or blood groups (biochemical polymorphism), in the morphological characteristics of chromosomes (chromosomal polymorphism), or in the DNA, by differences in the nucleotide sequence (DNA polymorphism). Nucleic acids and proteins have the property of appearing in different molecular forms or in multiple alleles, which may have implications in molecular pathologies.

The allelic variants of the coding sequences of genes in the DNA, due to normal processes of the cell function, promote the production of polymorphic codons and, with this, in alternate forms of proteins, generally without altering the function of the synthesized product, which allows differentiating a polymorphism from a mutation, which usually occurs at random. The polymorphism may happen in the promotor region of the gene, which influences the messenger RNA gene expression and, therefore, in the protein that it encodes (different phenotypes) or even identifies itself in untranslated regions (introns) so that there is no interpretation of its function, at least known so far. However, they are still sequences that allow differentiating individuals and species. Polymorphisms, like mutations, can be categorized by how they change the genetic code. Silent polymorphisms do not change the translation of the protein product when the sequence changes. As the nucleotide sequence changes, the codon that codes for the original amino acid changes to another codon that codes for the same amino acid or one with similar chemical properties. In addition, it considers that the neutral polymorphs, which are the ones that vary in their sequence in non-coding regions of the DNA, are also silent. Polymorphism is considered neutral if the presence or absence of the allele does not confer any advantage or disadvantage to the individual. In addition, a polymorphism may represent evolutionary advantages for a given population, such as conferring resistance to environmental conditions, according to the geographical area which they inhabit. The term polymorphic helps define genes or alleles. Even sequences that vary greatly are observed, such as in the alleles that determine the blood groups and molecules of human leukocyte antigens (HLA, human leukocyte antigen), constituting the major histocompatibility complex), which are considered highly polymorphic.

According to the fundamental principle of polymorphism in genetics, a population is polymorphic if a particular allele at a specific locus occurs with a frequency of more than 1%. This percentage indicates that these variants are familiar and do not occur by chance, *i.e.*, the least common allele cannot maintain its frequency simply by mutation. Polymorphism accumulates in populations until it becomes common between species; they are called genetic divergence. Thus, with time, a polymorphism could occur in many people and even be as common as a wild allele. Therefore, an allele that could initially be considered polymorphic may be the most common in a population.

The combination of sequences in the inherited alleles is unique for each offspring; thus, the analysis of their polymorphism allows us to obtain a definite pattern or profile unique for each organism, similar to the barcode of the supermarket products. It is called an individual genetic fingerprint due to its resemblance to fingerprints and distinct traits. Polymorphism is used as a genetic marker to identify or relate people since being inheritable, usually without being unchanged from parents to children, allows establishing a direct biological relationship since they share the same markers. To be able to carry out these marker studies, first of all, it is necessary to define the types of polymorphism genetic, as they can be classified according to differences in the structure, form of transmission, distribution, stability, and size. Since the study methods and identification of polymorphisms are diverse, and each option has advantages and disadvantages, using various types of polymorphisms that provide different information about their heredity and individuality allows comprehensive and valuable interpretation [2].

Types of Polymorphism

The characteristics of the sequences that make it possible to detect polymorphic sites are highly variable. Thus, it can differentiate two types of polymorphism: changes in a single nucleotide and deletions or insertions of a few or many base pairs.

According to this classification, the most commonly used polymorphic markers for genetic profiling are:

- Single nucleotide or single nucleotide polymorphisms (SNP).
- Restriction fragment length polymorphisms (RFLP).

- Variable number tandem repeats (VNTR).
- Short tandem repetitions (STR).
- Y chromosome Polymorphisms.
- Mitochondrial markers.
- Insertion-deletion (Indel).

Single Nucleotide Polymorphism

As the name implies, SNPs are variations in a single nucleotide or base pair in a given sequence. Deleting, inserting, and substituting a nitrogenous base in the expected nucleotide sequence can represent these polymorphisms. This type of polymorphism is persistent, and it has been described that finding an SNP in human DNA is possible every 1000 to 3000 bp in protein-coding sequences. The advantages of SNPs as biomarkers are their wide distribution throughout the human genome, their frequency, and their stability; moreover, due to their high frequency, many genetic diseases are caused by an SNP or related to identifying a specific SNP.

Restriction Fragment Length Polymorphism

This type of polymorphism is evidenced by restriction enzymes, which will cut specific DNA sequences. In this way, fragments are generated (restriction), which can hybridize to specific probes to be able to recognize them. Suppose an SNP affects the restriction site of an enzyme, a condition, or RFLP. When the nucleotide change alters the target sequence for a restriction enzyme, the polymorphism can be detected based on the variation it generates in the restriction pattern. In cases where the restriction site is recognized in both DNA strands, it will be considered a homozygote. Resolving the fragments in an electrophoretic run will result in a single band on the gel at a determined distance.

The fact that an individual is heterozygous means they have different alleles at a specific locus, by which identifying its sequence on a gel will be observed in more than one band: one for the fragment without restriction cut and others with the cut. In this case, the number of allelic variants is always two (C: short/NC: not short). It is handled as a function of identifying the restriction site by the enzyme used in digestion, *i.e.*, whether or not it cuts in the analyzed sequence. Instead, they can identify three genotypes: homozygous for C/C cut, heterozygous C/NC, and homozygous for not cutting NC/NC. In this way, individuals are identified with recognition of the restriction enzyme in a DNA fragment by segments of different lengths according to the generated restriction fragment.

Differentiating homozygotes from heterozygotes through this polymorphism allows us to define RFLPs as codominant; both alleles manifest without dominating one's expression on the other. However, even though the RFLPs are proof of heritable variation in DNA, this polymorphism is usually challenging to find (Fig. **1**). In any case, it only allows us to determine whether or not a polymorphic site is present in a specific sequence. Therefore, its use in indirect diagnosis is limited since they usually have fewer alleles.

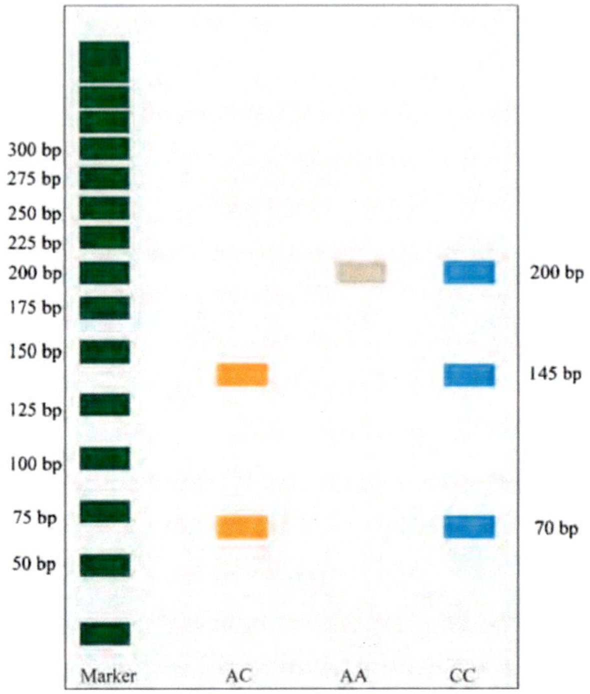

Fig. (1). TNFR1-383 SNP, identified by PCR-RFLP. Fragment digestion expected or the TNFR1 polymorphism. The banding pattern obtained from a gel is shown where digestion with the enzyme is evident. AA (135 and 64 bp), which identifies the homozygous for the restriction enzyme; AC (199, 135 and 64 bp), where a heterozygote is observed, which shows a band without a cut because there was no recognition by the enzyme, and another pair of bands showing the enzymatic cut by the recognition of the restriction site, and CC (199 bp), that shows the homozygote without a cut, that it does not present the recognition site for cleavage by the enzyme, and therefore both alleles are evidenced with a single band.

Polymorphism with Tandem Repeats

The VNTR and STR are loci that correspond to short sequences of DNA (one base pair or trio), block or tandem repeats a specified number of times. The length of these sequences can be from a few bases up to a few tens of nucleotides, and the number of repeats will determine the variability of different alleles at the same locus. Thus, the same chromosome regions with varying numbers of repeats can exist in the population. These polymorphisms are frequently located in intronic or non-coding regions, their length is variable, between 20 and 100 bp repeats, and their variability is high within the populations, allowing for their reliable use in identifying individuals. Because polymorphisms are heritable, a person may present VNTR inherited from his mother and father but could not present the VNTR that his father did not have. In this way, an individual's repeated sequences present a unique pattern. The more VNTR, the more distinctive and individualized profile or pattern obtained will be like fingerprints, only on a molecular scale.

Y Chromosome

The Y chromosome has genetic recombination with its homolog (X chromosome) only in about 1% of its sequence; therefore, the non-recombinant portion is of particular interest, as it contains sequences that are not found in other regions of the nuclear DNA, besides they are a purely paternal inheritance to sons. If the interest is to determine family relationships of paternal inheritance to compare relatives or male descendants, the most valuable polymorphisms would be specific Y chromosome sequences, which were identified by amplifying certain regions using specific primers.

The sequence of the heterochromatin of the Y chromosome is variable, and throughout the molecule, short repetitive sequences have been identified, such as AC repeats. In addition, it should be considered that the mutation rate for the Y chromosome is very low and has no negative implications for reproduction. These variations are beneficial for identifying relatives or tracing patrilineal descendants in the evolutionary studies of the human species; this has also served to determine a common origin or product of a single migration coming from Africa.

Mitochondrial Polymorphic Markers

The mitochondrion is a cell organelle that exists in multiple copies depending on the cell type. The mitochondria are exclusive of maternal inheritance; all descendants will have mitochondria from their mother in their cells. This is because, at fertilization, the egg preserves the mitochondria in the cytoplasm. On the other hand, the sperm's mitochondria, which are in the neck between the head and the tail, are taken out and do not become part of the embryo. This method allows identifying individuals according to the relatives by maternal line since all siblings, men, and women, will share the same information in their mitochondrial DNA (mDNA). The advantages of the analysis of the mRNA are the multiple mitochondria (1×103-1×104 copies) per cell, each containing several molecules of mDNA, characterized by being a circular double-stranded and have a small genome (16569 bp, which codes for 37 genes). In addition, this genome has a high rate of mutation (5 to 10 times more than nuclear DNA), which confers hypervariability.

This type of analysis is helpful, especially for evolutionary, anthropological, or forensic analysis, or when nuclear DNA is not easily accessible, it is deteriorated or exists in trace quantities. In order to conduct such an analysis, it is necessary to extract the DNA. For this, it is essential to amplify regions that are required to determine from the mDNA, in particular the hypervariable regions HV1 (16024-16035, 342 bp) and HV2 (73-340, 267 bp), located in the control region, which is the area that regulates the transcription of the molecule. These segments of mDNA are then sequenced and compared with samples from relatives coming from the maternal line by the type of mitochondrial inheritance to detect differences in the gene sequence [3].

Insertion-Deletion

Polymorphisms consisting of inserted or deleted bases are described. The sequences involved are small, generally up to 10 bp; however, it is possible that present sequences are up to 1000 to 1500 bp, as in the case of the insertion of transposons (transposition phenomenon, present in some genes that can move through the genome and reinsert into locations in another chromosomal region).

Random Sequence Polymorphisms

It is possible to analyze polymorphisms of randomly amplified DNA fragments or sequences (RAPD). In this case, 8 to 10 bp primers with a random sequence are used for amplification so that finding the sequences of complementarity between the initiator and the study sample can amplify a representative genome sample. The advantage of this technique is that no prior knowledge of the sequence or homologous probes is required, as in the RFLPs, which makes it an option as a screening test.

Electrophoresis analysis will show multiple bands that represent different DNA fragments. If there is homology between the initiator and the sample, the fragment will amplify and show up as a single band. Suppose there is no homology between the initiator and the sample. In that case, the fragment will not amplify because of a change in the sequence that would make the docking region of the initiator. With this method, nobody can tell the difference between heterozygotes and homozygotes who have the sequence because both bands would be on the gel. It is called a screening test because it only tells if the sequence is in the sample, not if it is on multiple alleles or just one.

Protein Polymorphisms

Protein polymorphism studies can be performed by identifying the amino acid sequence. This is especially useful for the differentiation of isoenzymes in the cell cytoplasm by detecting mutations in DNA, which would cause variations in the protein's primary structure without altering its action on the same substrate, but that would only affect their affinity for it. Electrophoresis is used to do this analysis because if the order of amino acids with different physiochemical properties were changed, the protein's final structure would be different, and their migration patterns in electrophoresis would be different because the charges and interactions between amino acids would be different.

This analysis helps identify polymorphisms in proteins having an enzymatic function. These variants are considered codominant polymorphisms, as they may have functional proteins with their counterparts. The variants may give a different metabolic response and cause homeostasis disturbance. It is necessary to consider the environmental impact; since it is minimal, it is easier to analyze the consequences of polymorphism on enzyme activity. For example, it is worth mentioning as the most metabolic enzymes of the cytochrome P-450 (CYP450) family, which are involved in transforming, neutralizing, and eliminating compounds, for example, medicines.

PATTERN ANALYSIS OR DNA FINGERPRINTING

The precise study methods allow us to carry out the analyses of genetic markers almost with absolute certainty. However, the study of polymorphisms requires appropriate technical and scientific knowledge and extensive experience in interpreting the results, provided they are reliable. The known and the most useful polymorphisms are generally point modifications of the DNA sequence. Genetic profiles are created through polymorphisms with markers or regions with patterns of defined repetition. The genetic profile, or DNA fingerprint, will determine an individual's genotype according to the genetic markers analyzed. The greater the number of markers analyzed, the greater the certainty for identifying the individual or confirming the filial relationship, carrying out studies of paternity, genetics, or forensic genetics or criminology, for the molecular identification of individuals. Polymorphisms can be linked directly or indirectly to specific pathologies or disease processes. They can also be used to show a link between polymorphism and a nearby defective gene through linkage disequilibrium studies. They can also be used as susceptibility markers to show the risk of developing diseases with multiple causes.

These modifications are studied mainly for genetic risk for various diseases. However, we also consider the sequence changes of chromosome fragments, like polymorphic portions on the heterochromatin of the Y chromosome or sequences of repeated insertions or deletions. The latter also serves for genetic diagnosis; they are most useful for family identification studies, affiliation, identification of individuals in accidents, criminology, and forensic science, among others. Identifying SNP is significant in biomedical research since it has allowed us to know and understand the disease process and develop treatment, control, and prevention strategies. Its clinical application includes, for example, the identification of the E4 genotype of the human apolipoprotein E gene (APOE), which is associated with lipid metabolism diseases and susceptibility to developing Alzheimer's disease.

The Cis 677 polymorphism makes the methyl tetrahydrofolate reductase (MTHFR) gene, which is involved in folate metabolism, stand out. Because it makes a less active thermolabile enzyme, this leads to low levels of plasma folate and elevated levels of homocysteine, which is linked to cardiovascular disease. Polymorphisms in drug-metabolizing enzymes cause differences in treatment response and its effects between patients since the biotransformation of each compound can vary, depending on the individual's genotype. The SNPs in these enzymes have served in the early identification of drugs' side effects, incompatibility or resistance to therapies, and variation in the efficacy of medicines. All this helps the doctor decide the appropriate treatment for each patient. This process is known as personalized medicine. Pharmacogenetics is the study of the CYP450 genes, including CYP2D6, which metabolizes a large number of drugs and has multiple allelic variants. Four phenotypes are defined according to their gene product's activity, whether they will be fast, functional, slow, or poor metabolizers (without enzymatic function), which has important clinical implications. The compound metabolizes so fast that it may lose efficiency. On the other hand, a rapid metabolizer (or ultrafast) may have more than two copies of the active allele, producing highly functional enzymes with accelerated enzymatic activity. The acetylation polymorphisms of genes are also important in metabolizing many compounds with arylamines, hydrazine, and carcinogens. N-acetyltransferase (NAT2) is an enzyme found in the cytosol that catalyzes the breakdown of these substances; the discovery of polymorphisms has been linked to impaired enzyme activity. This slow acetylating phenotype risks increased susceptibility to colorectal or bladder cancer.

For genetic diagnosis, the study of SNP-type polymorphisms makes it possible to identify allelic variants associated with the risk of disease, especially all multifactorial ones. For example, several SNPs have been associated with the predisposition to type 2 diabetes, including SNPs in the PPARG gene (peroxisome proliferator-activated receptor gamma or NR1C3, rs1801282 Pro12Ala, C/G), a nuclear receptor involved in glucose metabolism and the catabolism and storage of acids fatty. These determinations help establish prevention programs since the environmental component is essential, being multifactorial.

The ENPP1 gene (ectonucleotide pyrophosphatase/phosphodiesterase-1 or Lis121Gln, C/A) encodes for a class 2 membrane glycoprotein and has been associated with insulin resistance. In addition, the IL6 gene (interleukin 6) is a cytokine with diverse biological functions, including the immune response. In addition, variants have been found in many of the associated SNPs, such that the product gene of the modified sequence would have a high probability of conferring functional alterations. On the other hand, the typing of microbial genomes makes it possible to determine the distribution of microbial types in a specific population, its pathogenicity, and resistance to treatments or immunogenicity. This makes it possible to monitor populations at risk in advance, identify transmission patterns, and create new drugs or vaccines.

Pattern Detection Methods

Polymorphic or DNA fingerprinting to find specific sequences with a pattern has been determined to allow restriction enzyme cleavage or that can be hybridized for detection; several common approaches are used based on the homology of DNA sequences with specific probes. According to Chargaff's rules, the degree to which two DNA sequences complement or match is called homology. A complete complementarity between two DNA fragments demonstrates high homology, while a complementarity only in a small portion of base pairs shows low homology. Various methods can be used to detect polymorphisms, but the Southern blot is ideal for detecting polymorphisms in the DNA. Using specific probes to identify repeated sequences (VNTR or STR) locates hybridization regions homologous to a sequence of interest. Likewise, PCR with restriction enzymes and electrophoresis allows a finding of target sequences.

There is also a technique in which PCR amplifies polymorphic regions, including several polymorphisms (AFLP). This automated technique allows the study and creation of phylogenetic trees based on a comparative analysis of individual DNA samples. So, VNTRs can be quickly and inexpensively distinguished, making this method popular even in laboratories or countries with low resources. For STR analysis, STR loci, first treated with specific primers to amplify fragments containing short sequences (four bases repeated) with continuous repetitions, will be separated by gel electrophoresis to document the number of replicates and distinguish patterns of repetition that can be compared and associated with a standard. There must be a sequence of references to establish or rule out associations. For example, DNA samples obtained from a victim can be compared to the suspect's sample in a criminal investigation. STRs are common, shared by 5–20% of people, which forces the search for combinations of loci in multiplex reactions to achieve an acceptable method of discrimination and precise identification by always considering a base or reference sequence with which to make the comparison.

CONTROVERSIES IN THE ANALYSIS OF THE GENETIC FINGER-PRINT

Identifying polymorphisms to establish a genetic pattern or fingerprint is helpful in various fields of study. For example, forensic sciences allow the comparison of suspects through samples of blood, hair, saliva, or semen to determine responsibility or exoneration in criminal trials. It is also helpful for identifying human remains, missing relatives, evidence of paternity, identifying immigrants or undocumented persons, establishing family biological relationships to confirm, for example, the legal status of a minor, nationality, or origin, and determining the compatibility of organs in transplants, as well as defining the source or composition of food. Evolutionary studies can even be used for wild animal evolution.

Although this method of identification by the DNA fingerprint is considered equivalent to the code of bars of a commercial product, since each person has a specific genetic pattern, so far VNTRs only give evidence of probability (greater or lesser) that the genetic pattern is homologous with that of a particular person. However, this probability would be defined as a homology of 1 in 2 thousand million, which does not rule out the possibility that it is more common, the technique has been carried out inconsistently, and there are errors or false positives. This has been promptly addressed, and there are now universal standards and regulations for safety and precision in DNA fingerprint analysis, which minimize technical errors. Automation helps solve these problems of the processes and analyze combinations of rare polymorphisms (or uncommon among people). These allow the creation of patterns that increase the probability that the samples in comparison effectively match, in case of being identified, they come from the same person or the ones that correlate, this is the case of establishing family biological relationships, considering the variations due to differences in race or ethnicity. All of this heavily involves population genetics to develop relationships across racial lines.

In the future, DNA fingerprinting could be a valuable identification tool for all people being born, although there is still controversy regarding bioethical issues. So far, there are databases, especially of a judicial nature, that make it possible to identify suspects in criminal trials, which leaves out all the people who have not passed through such processes. It is proposed that requesting the fingerprint genetics of any person without any reason of character criminal law could be considered discriminatory. This would be easy to disprove if the test were done on everyone born at the same time since it would be a valid and reliable way to identify each person. However, the costs of doing it for each person are still very high, making it an impractical option for now and hardly possible. Another ethical implication refers to genetic diagnosis since analyzing polymorphisms associated with diseases produces results (positive or negative) that affect the family as a whole and not only the individual under study. On the other hand, the identification of risk factors or predisposition to various diseases would be beneficial for the proper prevention and treatment of diseases and to generate clear medical and socioeconomic benefits without neglecting, of course, the study and control of the environmental factors, with significant implications in the development and expression of a specific genetic profile [4].

GLOSSARY

Allele: The characteristics of a single copy of a specific gene or a single copy of a specific location on a chromosome.

Autosomal DNA: DNA found in chromosomes that are not sex chromosomes.

Chromosome: The biological structure by which hereditary information is physically transmitted from one generation to the next; located in the cell nucleus, it consists of a tightly coiled thread of DNA with associated proteins and RNA; the genes are arranged in linear order along the DNA.

DNA Profiling: The result of determining the relative positions of DNA sequences at several locations on the molecule; each person (except identical twins) has a unique DNA profile when used in the context of the CODIS database, which evaluates 13 specific DNA locations.

DNA Fingerprinting: Analyses of the lengths of the fragments reveal that when looking at multiple VNTRs within and between individuals, no two people have the same assortment of lengths, except identical twins; this technique became known to the public as "DNA fingerprinting" because of its powerful ability to discriminate between unrelated individuals.

Locus: The specific physical location of a gene on a chromosome; the plural form is loci.

Low Copy Number Analysis: The analysis of samples containing a small amount of DNA (approximately 30 cells or less); analysis of samples falling into this category often requires enhanced analysis methods to increase detection sensitivity.

Mitochondrial DNA (mDNA): DNA located in the mitochondria found in each cell of a body; sequencing of mitochondrial DNA can link individuals descended from a common female ancestor.

REVIEW QUESTIONS

1. Define the applications of DNA fingerprinting.
2. Discuss the probes used in forensic science.

REFERENCES

[1] Daniels M, Trowern A. Pharmacogenomics: SNPing away! Biochemist 2003; 25(1): 19-23. [http://dx.doi.org/10.1042/BIO02501019]

[2] Hartl DL. A primer of population genetics. Sinauer Associates, Inc. 1988.

[3] Wang H, Kim RA, Sun D, *et al.* Evaluation of the effects of 18 non-synonymous single-nucleotide polymorphisms of CYP450 2C19 on *in vitro* drug inhibition potential by a fluorescence-based high-throughput assay. Xenobiotica 2011; 41(9): 826-35. [http://dx.doi.org/10.3109/00498254.2011.582893] [PMID: 21692664]

[4] Willer CJ, Bonnycastle LL, Conneely KN, *et al.* Screening of 134 single nucleotide polymorphisms (SNPs) previously associated with type 2 diabetes replicates association with 12 SNPs in nine genes. Diabetes 2007; 56(1): 256-64. [http://dx.doi.org/10.2337/db06-0461] [PMID: 17192490]

CHAPTER 9

Application of Molecular Biology in Biotechnology

Hira Asghar[1,*]

[1] *Department of Science of Dental Material, Azra Naheed Dental College, Superior University, Lahore, Pakistan*

Abstract: Molecular biology is the study of molecular interactions and structures that govern cellular processes. Biotechnology is an interdisciplinary field that utilizes biological systems to create new products and technologies. The application of molecular biology in biotechnology has revolutionized the way we understand and manipulate biological systems. This technology has enabled the development of new diagnostic tools, therapies, and drugs. The ability to manipulate DNA and RNA sequences has also allowed the creation of genetically modified organisms with desirable traits, including crops and animals. In addition, molecular biology has facilitated the development of gene therapy, where defective genes are replaced or repaired, and vaccines, where specific antigens are identified and synthesized. Integrating molecular biology and biotechnology has opened up vast opportunities for research and innovation in many fields, including medicine, agriculture, and environmental science. As a result, the application of molecular biology in biotechnology holds great promise for the future of science and technology.

Keywords: Biotechnology, DNA sequencing, Genetic engineering, Gene therapy, Genetically modified organisms (GMOs), Molecular biology, Recombinant DNA technology, Synthetic biology.

INTRODUCTION

Since time immemorial, humankind has used living beings for its benefit. In different ways, sometimes by accident and sometimes with great ingenuity, procedures have been discovered to improve animal and plant varieties. The agricultural revolution, animal domestication, and wine, beer, and yoghurt production are but various manifestations of biotechnology. Biotechnology has been defined as the application of biological organisms, systems, and processes in manufacturing and service industries. However, modern biotechnology uses DNA techniques to improve living things with a view to their utilization. Biotechnology does not exist as a scientific discipline but rather as a wide variety of subject

* **Corresponding author Hira Asghar:** Department of Science of Dental Material, Azra Naheed Dental College, Superior University, Lahore, Pakistan; E-mail: hir_08@hotmail.com

Sanaullah Sajid, Sajjad ur Rahman, Shahid Mahmood, Shayan Bashir & Mudasser Habib (Eds.)

areas. Its fruits have grown from a tree whose roots are the biological sciences, particularly microbiology, genetics, molecular biology, and biochemistry.

This word was first used in the early 1920s by the Leeds City Council in the United Kingdom when its members founded an Institute of Biotechnology. However, biotechnological processes date back some 5,000 years ago, when the production of alcoholic beverages by fermentation was discovered. The ancient Egyptians went so far back in time that they even used mouldy bread as a poultice for healing infected wounds and introduced a service to detect pregnancy based on the effect of urine on the germination rate of wheat. The term biotechnology has only recently been introduced into popular vocabulary due to the great potential resulting from the application of molecular biology techniques. The advent of PCR has revolutionized all areas of scientific research to the point of making a multitude of processes that previously required manual and imprecise procedures. With it, the promise that biotechnology could become a commercial success has become a reality in just a few years. In addition, the relatively new *in vitro* genetic manipulation has expanded the number of products generated by microorganisms. It has provided new methods to increase the yield in obtaining existing ones.

Recombinant DNA technology has also significantly impacted the industry due to the unlimited possibility of separating a gene or DNA sequence from any source and transferring it to another organism so that it can replicate and be passed on to offspring. This sequence can be a gene that codes for a commercially helpful product, transferred to species in which it can be synthesized efficiently and economically, or a new trait for the host microorganism. The essential elements of the recombinant DNA technique are: 1) Obtaining specific DNA fragments (restriction enzymes). Restriction endonucleases are invaluable for fragmenting DNA at specific sites, as they recognize a particular sequence, which differs for different species and strains. 2) To reassociate two DNA molecules, covalent ligation or reassociation is necessary to reconstitute a phosphodiester bond; an enzyme called DNA ligase comes into action. 3) Mechanism to introduce recombinant DNA into the organism (transformation). Mechanism to ensure replication and identification of the recombinant molecule within the cell. If the transferred DNA is to be able to replicate, it must be bound to a molecule capable of multiplication in the host cell. A small molecule is required, which also possesses the replication capacity (*i.e.*, independent of the chromosome); these molecules are called vectors. In nature, there are two kinds of vectors: plasmids (extrachromosomal DNA molecules that are found as multiple copies in many bacterial species and are generally not necessary for cell viability, although they can carry some genes, such as those that confer antibiotic resistance), and bacterial viruses or phages, which can replicate inside a bacterium and give rise to many identical molecules. The bound molecules are introduced into host cells and selected by marker genes in the cloning vector. Within their sequence, cloning vectors contain signals that induce DNA replication.

Currently, with the introduction of PCR by Kary Mullis in 1983, restriction enzyme and sequencing, and hybridization techniques, the sequences of a single gene can be localized among an abundance of other sequences, cut at specific sites, and amplified exponentially to yield an analyzable quantity. PCR has increased the possibilities offered by DNA technology, as it allows the amplification of specific DNA sequences without the need for cloning in cells. This chapter aims to show how molecular biology has provided the primary impetus in the development of biotechnology and to describe how the potential of these techniques can be utilized in obtaining products or processes useful for man in some of the areas in which molecular biology has had the most significant impact, such as the pharmaceutical and agri-food industries [1].

HEALTH AND PHARMACEUTICAL INDUSTRIES

The emergence of genetic engineering and its impressive development to date has made it possible to initiate an in-depth analysis of the structure and mechanism of action of the gene program of a living being. The applications of this new knowledge have consolidated the origin of a new medicine that seeks to understand the molecular basis of health and disease. Today's therapy successfully uses genetic technology that allows the production of recombinant proteins necessary for diagnosing and treating genetic and infectious diseases and developing new vaccines. These methodologies have also affected the diagnosis and treatment of hereditary disorders. It will soon offer a new therapy to introduce healthy genes into cells with defective genes within our body.

The commercial application of recombinant DNA technology began in the late 1970s with the production of proteins as therapeutic agents. Several new therapeutic agents are based on genetically engineered antibodies, antisense technology, and other gene products and therapies through molecular biology techniques. However, therapeutic proteins have had the most significant impact on their use. In addition, their synthesis has been achieved in the quantities required for medicinal use. Typical examples are the transfer *to Escherichia coli* of the human insulin gene, used in the treatment of diabetes; the interferon genes for viral diseases and in combination with other agents in the treatment of different types of cancer; the human growth hormone messenger in the treatment of pituitary dwarfism; erythropoietin, which stimulates erythrocyte production and is used in the treatment of acute anaemia or related to treatments, and myeloid colony-stimulating factor, to complement anti-cancer therapy and in bone marrow transplants. As molecular biologists began the need for authentic and pure proteins, the demand for non-authentic derivatives began. With cloning, sequencing, and protein expression tools, the protein derivatives were prepared with altered amino acid sequences called muteins. These are proteins with new or enhanced biological activity, which studies certain substances' modes of action and physiological roles. Examples of muteins evaluated in clinical studies or disease models include tumour necrosis factor, renal cell carcinoma, and tumour growth factor.

On the other hand, the yeast Saccharomyces cerevisiae has been increasingly used in recent years to obtain heterologous proteins. *S. cerevisiae* was developed as the first eukaryotic expression system since, as a single-celled microorganism, it could be genetically manipulated by most techniques commonly used for *E. coli*. It is a suitable host for producing authentic eukaryotic proteins as a eukaryotic organism. *S. cerevisiae* is also safely suitable for producing recombinant pharmaceuticals because, unlike *E. coli*, it does not produce pyrogens or endotoxins.

A large group of heterologous proteins have been expressed in *S. cerevisiae*. Some have been produced commercially, such as the hepatitis B vaccine, tumour necrosis factor, and insulin. The *S. cerevisiae* expression systems are potent and versatile, as they have no limitations. It is often necessary to experiment with different methods before optimal expression, and even then, producing authentic glycosylated proteins may not be possible. Alternative yeast hosts, such as *Pichia pastoris*, are being developed to express heterologous genes with higher expression levels and greater efficiency in post-translational modifications (phosphorylation and glycosylation) than those obtained by *S. cerevisiae*.

Mammalian cell culture is more expensive than bacteria and yeast, and for reasons of economy, its use is limited to obtaining molecules and natural products from animal cells. The two most characteristic examples are monoclonal antibodies and interferons, and the value of these products justifies their high production cost. Recombinant DNA techniques have also been beneficial in identifying and validating new drugs, particularly in infectious diseases. Specific potentially important viral genes have been isolated from the viral genome, mutated to destroy their activity, and reinserted to see if the gene product is essential for pathogenicity. This technique has been applied to demonstrate which target genes are valid in AIDS therapy.

Although nucleic acid-based therapeutics have not yet been commercialized, the capacity of these systems is quite good. Antisense or oligonucleotide and gene therapy are the two main categories of future nucleic acid-based treatments. Oligonucleotide therapy involves inhibiting gene expression by hybridizing antisense oligonucleotides to a specific region of the "sense" mRNA or utilizing oligonucleotides acting on the gene itself that forms a DNA triplex when it binds to the DNA groove of the double-stranded DNA. This technology has been used to inhibit the expression of oncogenes and the growth of viruses in culture. Theoretically, any disease characterized by inappropriate gene overexpression is amenable to oligonucleotide therapy.

Gene therapy involves transferring an exogenous gene and a stable expression of the transferred DNA. It has been used to treat genetic deficiencies and infectious diseases and is now contemplated as a cancer therapy. In its simplest form, gene therapy can be applied to cells that can be removed from the body and reintroduced back into the body (ex vivo), such as blood cells. However, even in the most challenging cases, such as diseases in non-regenerable tissues and muscles, research is already underway to apply gene therapy. Presently, gene therapy procedures have been approved for use in diseases such as cystic fibrosis and some types of cancer [2].

FORENSIC MEDICINE

Determination of genetic fingerprints constitutes another application of molecular techniques, specifically PCR and restriction enzymes. These techniques are currently applied in medicine, forensics, and police investigations to identify individuals from biological samples, such as blood, semen, saliva, skin, and hair, found in cases of death, rape, and other types of aggression. The DNA profile allows the identification of such evidence with a degree of certainty previously considered impossible since it is possible to compare different DNA samples to check whether they belong to an individual or not or kinship between them. Similarly, the technique can provide evidentiary evidence in case of paternity claimed or unacknowledged [3].

VACCINES

Vaccination against bacterial and viral diseases has been one of the great successes in the history of human and veterinary medicine. The most significant example of the efficiency of vaccination is perhaps the eradication of smallpox. Despite the tremendous progress, contagious diseases are still a considerable problem for humankind. More than 10 million deaths worldwide each year are due to infectious diseases. Vaccines are the simplest and cheapest way to reduce the costs of infectious diseases. Genetic manipulation provides new techniques to address the problem in different ways. There are two classic strategies for vaccination. One is vaccination with killed pathogens or their subunits. The other uses live attenuated viruses or bacteria that do not cause disease but are derived from a pathogenic microorganism. The use of genetically engineered microorganisms represents an attractive alternative for the production of vaccines. The aim of producing a vaccine is to contain molecules with the same antigenic response without causing the harmful proliferation of the pathogen in the body. This is achieved using viruses that somehow transform into pathogens with the help of molecular biology techniques; cloning and expressing the genes of the viral envelope proteins in bacteria is possible using the bacterial product as a vaccine.

The most common vaccines for infectious diseases in humans are diphtheria, tetanus, pertussis, polio, measles, mumps, rubella, *etc*.; there are other vaccines for particular circumstances, *e.g*., cholera vaccine, yellow fever, Japanese encephalitis, hepatitis B, and so on. Some vaccines are satisfactory from the veterinary medicine point of view: rinderpest, foot-and-mouth disease, Newcastle, and Marek's disease. For some pathogens that are difficult to grow *in vitro*, such as the hepatitis B virus, the agent of human syphilis (*Treponema pallidum*), and the bacterium that causes leprosy (*Mycobacterium leprae*), recombinant DNA technology allows the transfer of genetic information from the pathogenic microorganism to other more manageable hosts such as *E. coli*, yeast or mammalian cells that make it possible to obtain live attenuated or inactivated vaccines [4].

FOOD INDUSTRY

In the food industry, bread and cheese are just two foods with specific genetic manipulations available to the consumer. Yeast has been used since ancient times in food fermentations in bread dough and the production of fermented beverages. Knowledge of genetics and biochemistry of yeasts has led to the development of new strains employing molecular techniques. In yeast, producing new strains with rapid growth and high yields has potential value. In bread making, new strains with fast development and good carbon dioxide production have been achieved. In the beverage industry, specific changes have been provided by cloning yeast in many enzymes for the production of lighter beers due to more efficient amylase activity, control of undesirable flavours due to the production of diacetyl and phenols, improved filterability due to the reduction of residual β-glucans, as well as improvements in storage quality through moisture control. In the same way, several improvements can be envisaged for dairy products, including the ability to produce stable acids and facilitate cheese making, the production of cheese flavour and texture enhancers, and accelerating ripening.

Some new strains are produced through non-recombinant techniques, such as *Lactobacillus casei* spp. Ramosum GG, with the ability to bind to cells through the human intestinal mucosa, stability to acids and bile salts, and *in vitro* production of an antimicrobial substance. Its usefulness lies in producing a fermented whey beverage and yoghurt-like product designed to promote human health by helping to control diarrhoea and constipation. Concerning foods that make use of lactic microorganisms produced by recombinant DNA techniques, we can mention cheeses that use the enzyme (responsible for milk coagulation), the main component of calf rennet, created by three different transgenic microorganisms, namely *Kluyveromyces lactis*, *Aspergillus niger* variant awamori and *E. coli* K-12. This enzyme is identical to that produced by calves. Chymosin made from transgenic organisms with respect to rennet can be collected faster, contain a higher concentration of active enzyme, and produce cheeses with controllable texture and aroma properties. Moreover, the incentive for its production lies in economics.

Many processes in the food industry may be amenable to enzyme utilization, representing a potential field in the genetic modification of cation of microorganism species to improve their stability, specificity, and production. A good number of enzymes used in food have been cloned. It is essential to mention that when the sources are derived from genetically modified microorganisms, their approval must be obtained before their use for commercial purposes, if possible. Many products include food additives, such as low-calorie sweeteners, flavour modifiers, vitamins, dyes, colourants, dehydrators, and nutritional supplements awaiting approval to enter the market. One example is a modified strain of *Acetobacter* capable of producing a new form of bacterial cellulose with a high capacity for hydration so that it may be an efficient slimming agent.

In food quality analysis, recombinant DNA techniques can identify cations and characterize and quantify pathogenic, spoilage, and food contaminants. Many DNA probes have been used to detect pathogens, such as *Salmonella, Listeria monocytogenes, Yersinia enterocolitica, Vibrio cholerae,* and certain viruses. Another technique that offers a faster and more sensitive method for detecting microorganisms in food is the genetically engineered strains capable of producing bioluminescence. Thanks to these methods, it is possible to quickly identify, characterize and quantify a pathogen present in a food. This information is handy as it makes it possible to decide the industrial destination of raw materials used in food processing and the affected foodstuffs from the sale [5].

PLANTS

Apart from providing food, plants are also an essential source of raw materials and valuable drugs. From the molecular point of view, the study of plants has been lagging behind other microorganisms. Recombinant DNA technology did not impact the area of plants as in health. This situation has changed in recent years. In recent years and considerable research, groups have detected the importance of applying these techniques to plant biology and the great potential of manipulating plants. The possibility of genetic engineering in plants depends, as in microorganisms, on introducing exogenous genetic material into plants and having it established and inherited from cell to cell. In case of plants, this is facilitated with the help of the microorganism *Agrobacterium tumefaciens*. This bacterium has made transgenic plants for millions of years by invading the plant and introducing a segment of DNA called a Ti-plasmid, developing a tumour. Another phenomenon facilitating plant engineering is plants' regenerative capacity since many plants can generate a complete organism from one cell (totipotent). Today, it is possible to introduce DNA into plants without the need for this bacterium.

In recent years, bio-ballistics have established mechanisms to transform and regenerate plants. This technique introduces DNA into the cells, passing through the cell wall using multiple shots with microscopic bullets that carry the genes and insert them into the plant tissue. The first tests on transgenic plants were carried out in 1986, and since then, there has been an impressive increase in their use. The rapid development of molecular biology techniques provides further advances in modifying agriculturally useful treatments in vegetable crops. It is now technically possible to identify, isolate, modify, transfer, and obtain the expression of a wide range of specific genes. There are many examples where a plant crop has been made an economically stable modification with a helpful phenotype. Genetic engineering of seeds used in food is a rapidly developing field. In theory, it is possible to obtain plants specifically engineered to possess the combinations of desirable qualities, such as disease resistance, stress tolerance, nitrogen-fixing ability, production of larger plants, improved storage properties and sensory attributes, acceleration, or retardation of ripening, and obtaining new sources of protein. Different breeding objectives exist for each crop, and a more specific set of needs may arise within the same species.

Molecular biology techniques can be used in several ways to facilitate crop improvement, enhance routine developmental processes, or increase the quality of existing varieties. The biochemical properties of crops can also be modified to improve quality concerning technological and industrial needs; for example, the quality of wheat for bread makes the quality of barley for brewing beer. Another objective pursued by plant engineering is related to resistance to insects, viruses, fungi, bacteria, and nematodes. There have also been rapid advances in molecular techniques to identify plant and seed pathogens and produce herbicide-resistant plants.

Molecular techniques are used in agricultural processes to reduce the losses caused by accelerated ripening due to ethylene production. The two enzymes involved in the ethylene production pathway have already been cloned. For example, transgenic mango and tomato crops can continue without rotting longer than other plants and are transported with less risk. A similar process can be used to prolong the transport and half-life of some flowers.

GLOSSARY

Adulterant: It is that substance or thing that is added to falsify something, material or immaterial, by mixing it or adding it in substitution to it; for example, contaminated food is food to which substances are added that do not correspond to its nature, in terms of its composition and definition (milk adulterated with vegetable fat).

Altering Agent: It modifies the determined characteristics and conditions.

Bioluminescence: Production of cold light by some organisms (bacteria, fungi, fireflies, abyssal fish, *etc.*). It has different functions: sexual attraction, recognition, prey attraction, and defence.

Sweetener: A substance with a sweet taste used as a substitute for sugar.

Endogenous: Originating or arising from within.

Exogenous: Originating or arising outside the body due to an external cause.

In vitro: Applied to biological processes when they are experimentally produced and isolated from the organism as a whole.

Oncogene: In biology and medicine, a gene derived from another gene that is part of a cell's normal genome and capable of converting that cell into a cancerous cell.

REVIEW QUESTIONS

1. What are the applications of cell biology in biotechnology?
2. What is the difference between biotechnology and biomedical engineering?
3. Is biotechnology better than molecular biology?
4. Describe the advantages and disadvantages of biotechnology.

REFERENCES

[1] Armendariz-Borunda J. Genomic medicine in Mexico. Applications of gene therapy for cirrhosis reversion. Ann Hepatol 2002; 1(4): 169-74. [http://dx.doi.org/10.1016/S1665-2681(19)32161-1] [PMID: 15280802]

[2] Hacein-Bey-Abina S, Le Deist F, Carlier F, *et al.* Sustained correction of X-linked severe combined

immunodeficiency by ex vivo gene therapy. N Engl J Med 2002; 346(16): 1185-93. [http://dx.doi.org/10.1056/NEJMoa012616] [PMID: 11961146]

[3] Fischer A, Hacein-Bey S, Cavazzana-Calvo M. Gene therapy of severe combined immunodeficiencies. Nat Rev Immunol 2002; 2(8): 615-21. [http://dx.doi.org/10.1038/nri859] [PMID: 12154380]

[4] Olivares EC, Hollis RP, Chalberg TW, Meuse L, Kay MA, Calos MP. Site-specific genomic integration produces therapeutic factor IX levels in mice. Nat Biotechnol 2002; 20(11): 1124-8. [http://dx.doi.org/10.1038/nbt753] [PMID: 12379870]

[5] Siller-López F, Sandoval A, Salgado S, *et al.* Treatment with human metalloproteinase-8 gene delivery ameliorates experimental rat liver cirrhosis. Gastroenterology 2004; 126(4): 1122-33. [http://dx.doi.org/10.1053/j.gastro.2003.12.045] [PMID: 15057751]

CHAPTER 10

Application of Molecular Biology in Gene Therapy

Maheen Shafiq[1,*]

[1] *School of Engineering and Built Environment, Griffith University, Nathan, Queensland, Australia*

Abstract: Gene therapy treats genetic and acquired diseases by introducing functional genes into cells to replace or correct defective genes. The field of molecular biology has played a significant role in the development of gene therapy, providing tools and techniques to manipulate and analyze genes and their expression. One of the main challenges in gene therapy is the efficient delivery of therapeutic genes to the target cells. Molecular biology has provided a range of vectors, such as viruses and plasmids, that can be used to deliver genes to cells and methods to modify these vectors to improve their efficacy and safety. Molecular biology has also contributed to the development of gene editing technologies, such as CRISPR-Cas9, which can be used to correct or modify genes at the genomic level. This approach can potentially treat genetic disorders by targeting the underlying genetic mutations. In addition, molecular biology has facilitated the development of methods to regulate gene expressions, such as gene silencing and RNA interference, which can be used to turn off genes that cause disease. Despite the progress made in gene therapy, many challenges remain to be addressed, including ensuring the safety and efficacy of gene therapy approaches. Continued advancements in molecular biology are critical for developing safe and effective gene therapies for treating genetic and acquired diseases.

Keywords: Functional genes, Genetic disorders, Gene editing, Gene expression, Defective genes, Gene silencing, Gene therapy, Molecular biology, Plasmids, Vectors, Viruses.

INTRODUCTION

Gene therapy is an area of genomic medicine that uses the arsenal available in molecular biology to introduce healthy copies of defective genes into specific cells of the organism and modify the course of the disease.

How does gene therapy work?

Researchers have developed several strategies to correct defective genes:

[*] **Corresponding author Maheen Shafiq:** School of Engineering and Built Environment, Griffith University, Nathan, Queensland, Australia; E-mail: maheen.shafiq@griffithuni.edu.au

Sanaullah Sajid, Sajjad ur Rahman, Shahid Mahmood, Shayan Bashir & Mudasser Habib (Eds.)

- Insert complementary DNA from a normal gene into the genome to replace the non-functional gene.
- Exchanging an abnormal gene for a normal one by homologous recombination.
- Repairing the abnormal gene by selective reverse mutation returns the gene to its normal function.
- The regulation of a gene can be altered by switching it on or off.

What is gene therapy?

Gene therapy studies the insertion of a “normal” gene to replace an “abnormal” gene (the gene that causes the disease). The carrier molecule is a vector to deliver the therapeutic gene to the patient's target cells.

What is used as a vector in gene therapy?

The most commonly used vector is a virus genetically modified to carry normal human DNA and cannot replicate autonomously. Viruses have naturally evolved a way to encapsulate and deliver their genes to human cells but in a pathogenic form. Scientists have utilized this ability and, after manipulating the virus genome, have inserted therapeutic genes to correct the disease and control the viral vector's replication.

What is the current status of gene therapy research?

The U.S. Food and Drug Administration (FDA) is responsible for approving any products that go on sale and are used for human gene therapy. Much of the gene therapy is in the experimental phase, and has not demonstrated much success in clinical studies. The first clinical study of gene therapy was conducted in 1990. The importance of gene therapy lies mainly in making these advances known to students of medicine and healthcare so they learn about this new methodology.

Gene therapy is a process that involves the introduction of nucleic acid into eukaryotic cells, excluding germ cells, with the aim of modifying the progression of a medical condition or rectifying a metabolic or genetic abnormality. In its early stages, gene therapy was primarily developed to address the specific intention of addressing monogenic illnesses, which are characterized by a mutation or deficiency in a single gene. However, it is worth noting that a solitary gene has the potential to be regarded as a novel “pharmacological agent” for the treatment of numerous illnesses. In the last decade, the concept of gene therapy has transitioned from a theoretical notion to a practical application, as evidenced by the implementation of 835 clinical procedures globally, involving a total of 5,625 patients. Notably, most of these procedures are being conducted within the United States. The early success of this relatively new sector from clinical and preclinical methodologies has been achieved, despite expectations being exceeded. Despite being a very nascent field, it merits attention due to its significant progress facilitated by recent technological advancements, enabling its application in treating attention due to its significant progress facilitated by recent technology advancements, enabling its application in the treatment of many ailments. The above observation has been demonstrated through the encouraging outcomes observed in individuals afflicted with severe X-linked combined immunodeficiency syndrome (SCID) and hemophilia B.

Nevertheless, it is crucial to acknowledge that several obstacles exist that impede the efficacy of the therapeutic intervention. The primary challenge that has proven to be particularly tough is the lack of efficient means to transport the "therapeutic gene" to a specific cell or tissue, resulting in the inadequate production of the desired gene product, typically a protein, which is necessary to correct the condition. This phenomenon can be attributed, in part, to the development of intricate mechanisms within cells and organs that effectively hinder the internal accumulation of the outside genetic material. The predominant emphasis in the advancement of gene therapy systems for human application has been directed towards somatic cells, mostly due to the substantial apprehension surrounding the potential introduction of foreign genetic material into germ cells. Nevertheless, recent research has investigated the potential of gene transfer to the animal uterus as a means to rectify defects in the developing embryo. The aforementioned facts necessitate additional analysis, although undeniably expand the scope of available treatment interventions. The methods employed for the therapeutic gene transfer of genes into human cells can be categorized into three overarching strategies: *ex vivo*, *in vivo*, and *situ*. One example of *ex vivo* techniques involves the extraction of leukocytes primarily from the patient, followed by their cultivation and stimulation in a controlled environment. Subsequently, genetic material is inserted into the cells, which are then sustained in culture. Once the desired gene is properly expressed, these cells are introduced into the diseased host through injection. In contrast to *in vivo* methodologies, the aforementioned multi-step procedure is circumvented with the direct administration of the therapeutic gene into the bloodstream of the afflicted host, hence facilitating targeted delivery to the specific organ. The process of in *situ* gene delivery involves the direct injection of a specific gene into a suitable vector, such as in solid tumors, such as prostate cancer [1].

THE LIVER AS A TARGET ORGAN

There are a number of qualities present in the liver that make it an appealing target for gene therapy. Because it is an organ in which critical metabolic pathways are carried out, the liver is susceptible to various metabolic conditions present at birth. The liver can carry out the role of a secretory gland and produce therapeutic proteins due to its unique position in relation to the circulation of the bloodstream. Finally, because of the unusual shape of the hepatic endothelium, the hepatic parenchyma is accessible to molecules in the bloodstream that are as large as DNA fragments or recombinant viruses.

In the last decade, there has been evidence of several procedures of transferring genes to the liver parenchyma in normal or diseased livers. Different investigators have used simple methods, such as injecting plasmid DNA, naked or linked with different ligands, by injecting recombinant viral vectors into the bloodstream, which are released precisely in the hepatic lobules. Thus, scientists repeatedly tried to achieve a level of transgene expression that is productive and stable in many hepatocytes.

THERAPEUTIC STRATEGIES FOR HEPATIC FIBROSIS

Antifibrotic treatments that are effective in theory are required to fulfill a number of essential requirements in order to be considered effective. To begin, each treatment must have a valid biological basis and a potentially beneficial pharmacological treatment, keeping in mind that the quantities of active chemicals must reach the liver in high concentrations while producing few extrahepatic effects. This is a prerequisite for any therapy. Lastly, it is essential for a therapeutic agent to maintain its presence in the microenvironment of the target tissue for an extended period of time. Regrettably, there is not yet a treatment that can be considered definitive for cirrhosis of the liver. However, recent discoveries on the underlying pathophysiology of liver fibrosis and the role that activated hepatic stellate cells play offer hope for developing therapies [2].

ADVANCES IN VECTOR GENE TECHNOLOGY

The lack of efficient vector systems to deliver a specific gene in a targeted manner is the main obstacle to the growth of gene therapy. A desirable vector can multiply to high titers in the target cells and remain there without causing any immunological or toxic consequences. Unfortunately, this ideal vector has not yet been developed, and it is doubtful that a universally suitable vector will soon be available. However, researchers in the field of gene therapy around the world have been reporting some promising advances in the field. Indeed, this biotech field is advancing quickly. Retroviruses continue to be extensively explored because of their unusual capacity to integrate into the host cell's chromosomes and being non-immunogenic.

Many retroviruses, including oncoviruses, lentiviruses, and foamy retroviruses, have been studied for their potential to treat hematopoietic disorders due to their diverse integration capabilities into progenitor cells. The treatment of numerous children with severe combined immunodeficiency syndrome (SCID-XI) is still the most prominent example of human gene therapy's success (to date). However, lentiviral vectors are improving in safety and convenience. Preparations with high recombinant titers and undetectable competent replication events can now be produced using inducible cell lines with packable multi-component. Lentiviral methods have improved their cloning capability, especially in adult and pediatric hematopoietic progenitor cells, and show high effectiveness for integration into nondividing cells. The capacity of foamy viruses to rapidly integrate into hematopoietic progenitor cells and clone any retroviral subfamily has piqued the interest of researchers.

The use of adeno-associated viruses (AAVs) as vectors is rapidly increasing. Very few immune reactions occur when these vectors are grown to high titers. New evidence from the Kay and High labs suggests that AAVs can efficiently express a gene in the liver, raising the prospect of future treatment for hemophilia. Muscle AAV delivery in an early trial resulted in modest increases in circulating factor IX. The AAV delivery of cloned genes to the liver is an active research area. Despite claims that AAVs can circumvent the cytotoxic T-cell response entirely, it has become evident that certain transgene combinations and promoters might result in severe complications. It will be necessary to get through this obstacle eventually. The prevalence of adenoviruses has fluctuated over time. Their employment of adenoviruses raises safety concerns in light of the 2001 clinical trial failures. However, the general consensus on using these vectors is that they are helpful and improve daily. Several research institutes throughout the world have adopted the use of adenoviral vectors in vector engineering. Adenoviral vectors come in a wide variety of serotypes, which has proved crucial to successfully targeting specific cell types by using modified serotypes and target surface proteins (acting as ligands). Adenovirus can develop into being extraordinarily high titers and overcome most immunological issues if its genome is completely deleted or empty. However, it has been demonstrated that some pathological changes, combined with transgenic promoter regions, might elicit lingering immunological problems.

Leaf Huang and colleagues showed that naked plasmids, the simplest vectors, can sometimes achieve transduction levels comparable to those of viruses. Virus delivery systems enhance the ability to translate in the specific target cell compared to non-viral technologies. This advantage, however, is threatened by rising worries about the safety of certain viruses and the great capability of delivering non-viral targeting vectors (mainly by conjunction with ligand molecules). It is tough to say what further gains might be expected. Nevertheless, human gene therapy would undoubtedly be brought closer to reality if vector technology were rapidly developed. In addition, each gene therapist's specific requirements will dictate which vectors are best-suited to transport their chosen therapeutic genes. As previously mentioned, the engineering and design of a foolproof universal vector have not yet been implemented [3].

Gene Therapy in Liver Fibrosis

Several recent scholarly articles have employed gene therapy to mitigate the effects of cirrhosis in diverse experimental models induced by distinct etiological factors. Consequently, the telomerase gene has been targeted to suppress and protect against CCl4-induced liver fibrosis in mice. Moreover, an alternative research team has documented that the translocation of neuronal nitric oxide synthase to the liver in rats with cirrhosis improves the condition of portal hypertension. Uesugi et al. conducted a notable study whereby they posited that the inhibition of NF-κB, achieved through the administration of its repressor IκB, effectively mitigates initial alcohol-induced damage in rats. Furthermore, the utilization of an adenoviral vector was employed to encode for hyper-IL-6, a modified cytosine super agonist comprising the human IL-6 connected to the IL-6 receptor *via* a flexible peptide chain. This approach successfully preserved hepatic function, impeded the advancement of necrosis, and stimulated regeneration in a murine model of acute liver injury induced by D-galactosamine administration.

In light of efforts to mitigate experimental chronic liver diseases, it is increasingly apparent that optimal strategies for managing cirrhosis should encompass measures aimed at preventing pathogenic progression. These measures may involve stimulating liver cell proliferation, facilitating the development of new hepatic vessels, and promoting the reorganization of liver architecture. One noteworthy observation is the transduction capacity of adenoviral vectors, namely those belonging to the first, second, and third generations, in cirrhotic livers. It is worth noting that this issue had not been effectively addressed in the context of murine livers. One notable benefit of the liver is its preferential selection as the primary target for the adenovirus, resulting in a higher proportion of viral particles being internalized into hepatocytes rather than non-parenchymal cells. The adenoviral vectors exhibit a preferential delivery of the therapeutic gene to the liver following injection *via* the systemic route. The targeted distribution of drugs to certain organs will result in a decrease in overall toxicity.

However, a notable drawback of numerous adenoviral vectors is the transient nature of gene expression and the related protein's functional activity. Previous studies have documented varying expression durations ranging from two to three weeks up to two to three months. The quality and titer of individual batches of adenoviral vectors mostly influence the specific expression length. Nevertheless, a potential solution to this issue appears promising in the foreseeable future since upcoming vectors of the next generation, commonly referred to as "gutted" vectors, are on the verge of being introduced. In contrast, the fusion of liposomes with various molecules is characterized by a high level of specificity, whereas the current shipment lacks such specificity.

The *in vivo* experiments used carbon tetrachloride (CCl_4) as the experimental model. The experimental study used the bile duct ligation model to assess the effectiveness of transduction. A dosage of 3 × 1011 viral particles/kg of the adenoviral vector containing the gene encoding matrix metalloprotease 8 (MMP8) was administered for this purpose. The justification for employing MMP8 was to facilitate the localized breakdown of extracellular matrix proteins, facilitate the release of hepatic growth factor (HGF), and provide vacant areas for the proliferation of liver cells. In contrast, an alternative therapeutic gene known as human plasminogen urokinase (HuPA) was employed, resulting in a significant reduction of 85% in hepatic fibrosis by day ten after administering the adenovirus. The introduction of adenovirus containing the HuPA gene resulted in a near-total resolution of periportal and centrilobular fibrosis, resulting in an 85% decrease in hepatic fibrosis by day 10. This reduction in fibrosis was found to be associated with an upregulation of MMP-2 expression. Furthermore, it was shown that livers affected by cirrhosis exhibited a significant presence of α-smooth muscle actin (α-SMA) positive cells. Furthermore, it was shown that Ad-HuPA-treated rats exhibited an augmentation in liver regeneration, alongside the resolution of fibrosis. Liver sections exhibited a notable augmentation in mitotic figures, binucleated hepatocytes, and cells expressing nuclear proliferation antigens (PCNA) over a span of two days following Ad-HuPA injection. Prior to implementing these biotechnological procedures on human patients, it is imperative to gather relevant data [4].

Antisense Therapy

This is another methodological strategy, where the diseases characterized by an abnormal protein or the overproduction production of a normal protein can potentially be treated by selectively blocking the gene(s) or genes that cause this disease. When a gene is expressed, the DNA base sequence is transcribed into a single RNA strand and then translated into a protein. The main idea of antisense therapy is to prevent the expression of the gene or protein RNA; that is, to interrupt the flow of genetic information using synthetic oligonucleotides that target specific mRNA sequences [5]. Since 1977, some researchers have used antisense therapy to inhibit DNA expression, at least in cell-free systems. Since then, the idea of regulating DNA expression in cell-free systems has emerged, including using bacteria and mammalian cells. In addition, researchers have tried specific sequences to silence some target genes.

There are two possible mechanisms by which the binding of a complementary oligonucleotide can block mRNA translation.

- Base-specific hybridization; prevents access to the mRNA by the cell's translation machinery (arrest hybridization).
- Cleavage of the RNA-DNA hybrid by the enzyme ribonuclease H.

Antisense molecules are usually small segments of DNA or RNA that exhibit specific characteristics:

1. They are a synthetic single-stranded polymer with a 3′-5′ sense.
2. They are generally made of DNA, although sometimes they can be made of RNA.
3. They have a specific sequence that, by complementarity, binds to an mRNA.

Antisense therapy attempts to interrupt the gene's expression in one of the two stages of protein processing. The first phase is transcription; this phase refers to encoding a gene into a messenger RNA that leaves the cell nucleus. The idea is to design an antisense oligonucleotide capable of inhibiting the specific growth of the DNA double helix so that a triple helix is created within a gene or in one of its promoter elements, which prevents the coding of the gene from being read. The other phase that can be interrupted is translation, where the messenger acts as a recorded instruction that enables the cell in the ribosomes to synthesize a protein. The idea is to block the messenger directly. If an antisense strand can bind and hybridize to a specific messenger, the resulting DNA-RNA duplex strand would prevent the ribosomes from reading the message. A slightly different idea, but also related to this blocking of the message, is related to the presence of ribozymes, in which the breakage of the RNA-RNA hybrid by the enzyme ribonuclease H. Ribozymes are RNA enzymes that synthesize with a specific target sequence, but as part of the RNA structure, are capable of self-catalysis. The cleavage site is on the target messenger RNA.

Whatever the mechanism, the goal is to disrupt the mRNA in a living cell; this is more complicated than the theoretical design alone tested in cell culture since the antisense strand has to reach the target cell by clearing a series of barriers. Firstly, it must have protection to be resistant to circulating nucleases. Thus, its design must give it protection of days rather than hours to reach the specific site that will target the action of this antisense therapy. Another critical point is that the oligonucleotide must enter the target cell; naked nucleic acids have a strong negative charge that makes the molecule soluble in water but poorly soluble in lipids, which impedes its ability to cross the plasma membrane. The cell also possesses different types of receptors that facilitate the introduction of molecules into the cell (Fig. **1**). Another idea is to associate the oligonucleotide with a carrier; at this point, liposomes are functional molecules for this purpose. The presence of endosomes and lysosomes, which are responsible for degrading small circulating fragments, must also be considered. Once it can clear these barriers and enter the nucleus through a nuclear pore, it must find its target. Suppose the gene or messenger is not linear enough to hybridize with the target sequence it designed. In that case, the success of achieving its inhibition is far from being achieved: in fact, both DNA and RNA are folded, and they are also folded and are also covered with regulatory proteins, which can hide the target sequence characteristics.

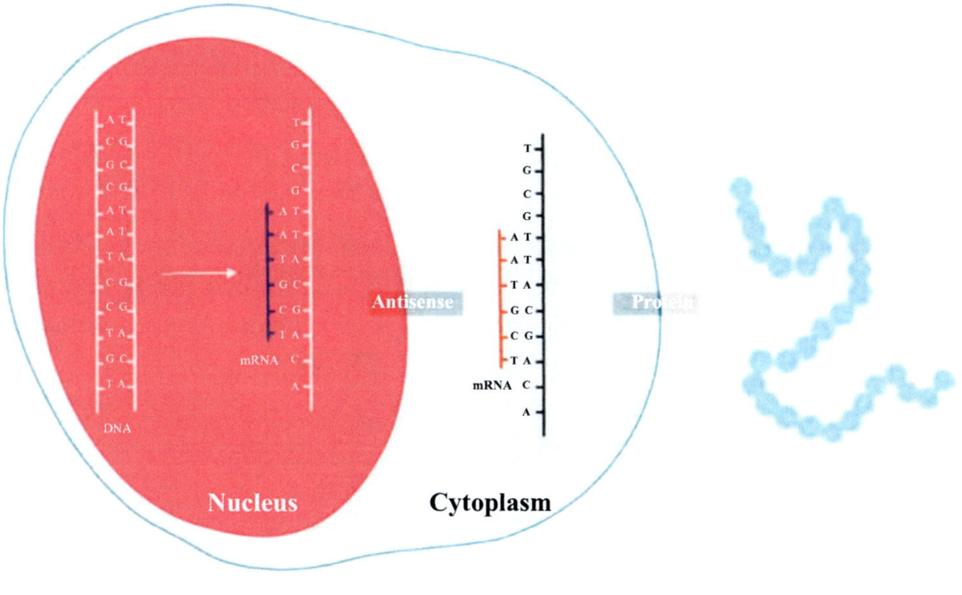

Fig. (1). Cell's translation machinery.

CONSIDERATIONS AND NEW STRATEGIES

Another important point is raised when it deactivates the message, the antisense DNA strand will continue to participate in destroying messengers. Likewise, the ribozyme will continue to kill the messenger; however, both systems are degraded. In this case, the important thing is that the inhibition of the message can be maintained for a longer time. DNA-DNA duplexes are weaker than RNA-RNA duplexes, and attempts have been made to design DNA more closely resembles RNA.

In cases of translocations, the ideal is to target the new oncogenic protein resulting from the fusion of two non-oncogenic sequences. The problem in these cases is that the breakpoint appears to be inaccessible. Another point is that after this exchange of sequences takes place at longer times, each stem cell divides and gives rise to a daughter cell that possesses the change, propagating this defect very fast in the genome.

GLOSSARY

Gene therapy: The aforementioned medical method involves treating or preventing disease by correcting the fundamental genetic anomaly. Gene therapy approaches enable researchers to address disorders by modifying an individual's genetic composition instead of relying on conventional research instruments or pharmaceutical interventions.

Complementary DNA: Double-stranded DNA with matching chemical sequences on each strand is called "complementary DNA" or "cDNA."

Plasmid: A cell's tiny, extrachromosomal DNA molecule can replicate independently. Plasmids are small, circular, double-stranded DNA molecules in bacteria, archaea, and eukaryotic species.

Vector: Any particle (such as a plasmid, Cosmid, or Lambda phage) can be used as a vector in molecular cloning to introduce a foreign nucleic sequence (often DNA) into a recipient cell for replication and/or expression.

REVIEW QUESTIONS

1. What are the risks associated with gene therapy?
2. Discuss the ethical issues related to gene therapy.
3. Gene therapy: a treatment for the rich. How accurate is this statement?

REFERENCES

[1] Armendariz-Borunda J. Genomic medicine in Mexico. Applications of gene therapy for cirrhosis reversion. Ann Hepatol 2002; 1(4): 169-74. [http://dx.doi.org/10.1016/S1665-2681(19)32161-1] [PMID: 15280802]

[2] Hacein-Bey-Abina S, Le Deist F, Carlier F, *et al.* Sustained correction of X-linked severe combined immunodeficiency by *ex vivo* gene therapy. N Engl J Med 2002; 346(16): 1185-93. [http://dx.doi.org/10.1056/NEJMoa012616] [PMID: 11961146]

[3] Fischer A, Hacein-Bey S, Cavazzana-Calvo M. Gene therapy of severe combined immunodeficiencies. Nat Rev Immunol 2002; 2(8): 615-21. [http://dx.doi.org/10.1038/nri859] [PMID: 12154380]

[4] Olivares EC, Hollis RP, Chalberg TW, Meuse L, Kay MA, Calos MP. Site-specific genomic integration produces therapeutic Factor IX levels in mice. Nat Biotechnol 2002; 20(11): 1124-8.

[http://dx.doi.org/10.1038/nbt753] [PMID: 12379870]

[5] Siller-López F, Sandoval A, Salgado S, *et al.* Treatment with human metalloproteinase-8 gene delivery ameliorates experimental rat liver cirrhosis. Gastroenterology 2004; 126(4): 1122-33. [http://dx.doi.org/10.1053/j.gastro.2003.12.045] [PMID: 15057751]

CHAPTER 11

Molecular Biology of Sports

Sanaullah Sajid[1,*]

[1] *Institute of Microbiology, University of Agriculture, Faisalabad, Pakistan*

Abstract: Molecular biology of sports is a rapidly evolving field that investigates the relationship between genetic and molecular factors and athletic performance. Research in this area aims to identify genes and molecules influencing physical traits such as strength, endurance, and speed. By understanding the underlying biology of athletic performance, scientists can develop new approaches to improve athletic training, injury prevention, and rehabilitation. The study of molecular biology in sports also provides insights into the relationship between genetics and lifestyle factors, such as diet and exercise, that affect overall health and well-being. This abstract will provide an overview of the current state of research in molecular biology of sports, including recent advances in genomics, transcriptomics, proteomics, and metabolomics, and how these techniques are applied to sports science.

Keywords: Athletic performance, Endurance, Health, Metabolomics, Proteomics, Physical traits, Strength, Transcriptomics, Well-being.

INTRODUCTION

Physical activity is considered any bodily movement carried out by human beings during a specific period, either as part of their work activity or in their leisure moments; this movement considerably increases energy consumption and resting metabolic levels; that is, physical activity induces calorie consumption. Daily physical activity benefits the health, cardiovascular, musculoskeletal, and immunological systems, *etc*. Physical activities integrated into everyday life, such as walking, bicycling, climbing stairs, doing housework, or simply making purchases, are beneficial, especially if they are frequently made. However, it is best practice to program physical exercise or sports at moderate intensity. Sport is any form of physical activity that aims to express or improve psychophysical attitudes, develop social relationships, or achieve sporting results at any level. The correct performance of a sport requires training. Training systematically plans and implements the necessary measures (contents and methods) to obtain specific persistent and long-lasting physical or psychological effects through physical

* **Corresponding author Sanaullah Sajid:** Institute of Microbiology, University of Agriculture, Faisalabad, Pakistan; E-mail: sanaullahsajid@gmail.com

Sanaullah Sajid, Sajjad ur Rahman, Shahid Mahmood, Shayan Bashir & Mudasser Habib (Eds.)

activity. Exercise requires body movements, which are achieved through skeletal muscle movements. The muscle cell can only obtain chemical energy through adenosine triphosphate (ATP). Therefore macronutrients must give up the energy of their chemical bonds through various biochemical processes. For this to happen, different metabolic pathways provide energy [1].

ENERGY PATHWAYS DURING PHYSICAL ACTIVITY

The primary energy for the contraction of muscle fibers is obtained from the breakdown of ATP into adenosine diphosphate (ADP) and adenosine monophosphate (AMP). This reaction is carried out by the enzyme myosin ATPase; However, the amount of ATP available in the muscle only provides power for 1 to 2 seconds. If muscle contractions are prolonged for longer, ATP has to be supplied through secondary energy sources, divided into anaerobic and aerobic energy pathways. Anaerobic efforts have a maximum duration of two minutes, while endurance efforts last more than 10 minutes, and 80% of the energy source comes from oxidative pathways, called the aerobic pathway, and its primary energy source is carbohydrates [2].

Anaerobic A-lactic Pathway: Phosphocreatine

The first mechanism of secondary energy pathways is the breakdown of the creatine phosphate (PCr) molecule into creatine: ATP + PCr $\rightarrow$ ATP + Cr. In muscle, PCr is reserved in small amounts; in untrained people, the energy from this metabolic pathway lasts up to six seconds; in high-performance athletes, it lasts 12 to 20 seconds. This energy pathway does not produce lactate, called the anaerobic a-lactic pathway. This pathway is activated during athletics races of 100m, the start of swimming, the long and triple jump, or javelin and discus throws, *etc.*

Anaerobic Pathway: Anaerobic Glycolysis

If the energy demand exceeds six to 20 seconds, the PCr runs out; from that point, energy is obtained from muscle glycogen through anaerobic glycolysis. During this metabolic pathway, the energy of the glucose molecule as a metabolite is generated by two pyruvate molecules. Then, in aerobic glycolysis, pyruvate can be transformed into acetyl coenzyme (CoA) by oxidative decarboxylation through a series of reactions catalyzed by pyruvate dehydrogenase, located in the mitochondrial matrix, and Acetyl-CoA enters the Krebs cycle to generate more energy. When the energy demand exceeds oxygen availability in the blood, pyruvate dehydrogenase fails to convert pyruvate to acetyl-CoA. Fast enough, it builds up, inhibiting glycolysis and reducing ATP production. Under these conditions, the enzyme lactate dehydrogenase (LDH) reduces pyruvate to lactate by the following reaction:

Pyruvate + NADH + H → lactate + NAD

The production of lactate oxidizes the NADH + H to regenerate nicotinamide adenine dinucleotide (NAD), which is necessary for glycolysis and to continue the production of ATP. In the anaerobic breakdown of glycogen, lactate is produced in concentrations ranging from 4 to 20 mmol/L of blood; accumulations greater than 6 to 8 mmol/L prevent efficient sports performance. Anaerobic glycolysis reaches its maximum level after about 45 seconds, and it lasts about two minutes; therefore, the duration of lactic anaerobic efforts is limited by muscle acidification (due to the depletion of the shock absorbing systems of the muscle that try to reduce the concentration of hydrogen cations released in the hydrolysis of ATP) and due to the lack of muscle glycogen. These efforts can be appreciated in the 400 m race and the 100 m swimming. On the other hand, the produced lactate leaves the cell muscle and reaches the liver *via* the bloodstream, where it is transformed back into glucose by gluconeogenesis, known as the Cori cycle.

Aerobic Pathway

This pathway can be divided into long-term physical exercise duration and short and medium-duration physical exercises. In long-term physical activities, energy comes mainly from lipids. Here the speed of energy production is low. Since fat is an abundant energy reserve, it allows for maintaining physical efforts for long periods, as in the case of marathons, triathlons, marches, and bicycle races. On the other hand, the energy necessary for short or medium-form aerobic exercises is obtained mainly from carbohydrates, whose combustion is faster than that of fats; at a particular time, the reservations of carbohydrates are depleted. In this exercise, lactate concentrations may appear in 2 to 4 mmol/L blood. This energy source is used in 400 m swimming events and 500 and 1000 m track races.

Aerobic Pathway: Aerobic Glycolysis

After approximately two minutes, the energy is obtained by the aerobic oxidation of the rest of the glycogen stores; acetyl−CoA is converted, together with oxalic acid (oxaloacetate), in citric acid, in each Krebs cycle which will generate one molecule of oxaloacetate to two CO_2, guanosine triphosphate (GTP) and high electrons potential in the molecules of NADH and $FADH_2$. The net balance of the cycle is:

Acetyl−CoA + 3NAD+ + FAD + GDP + Pi + $2H_2O$ → CoA−SH + 3(NADH + H+) + $FADH_2$ + GTP + $2CO_2$

This process is related to optimal energy use since the complete reduction of 1 mol of glucose recovers 38 ATP molecules. This energy source usually lasts about 30 minutes in untrained people and 60 to 90 minutes in trained people. After this period, muscle and liver glycogen stores are generally depleted.

Aerobic Pathway: Aerobic Lipolysis

The energy required to maintain ATP levels in the muscle during constant activity can be obtained from exogenous carbohydrates from food, as usual in cycling races, triathlons, marathons, *etc*. If, for any reason, this supply cannot be carried out during a high-intensity exercise, the necessary energy will be provided *via* hepatic glycogenolysis or by oxidation of amino acids. Thus, if the effort continues, even if this supply cannot be carried out during a high-intensity exercise, the necessary energy will be obtained from the aerobic oxidation of triglycerides, converting them into glycerin and free fatty acids. Aerobic lipolysis of 1 mol of triglycerides produces a power of 130 moles of ATP; however, obtaining energy is slow.

Types of Muscle Fibers and their Influence on Sports Performance

Sports performance is divided into slow *versus* fast and resistance vs. power and highly depends on cardiorespiratory functions, hypertrophy, and types of skeletal muscle fiber. Three types of skeletal muscle fibers are known: red, white, and intermediate. The functional difference of the fibers depends on the behavior of the myokine ATPase since those of red muscle use the oxidative pathway. In contrast, white muscle uses the anaerobic glycolytic pathway. Fibers are abundant in red muscles; their diameter is small and contains a large amount of myoglobin, so numerous mitochondria are located and accumulate below the sarcolemma. The fibers of white muscles are greater in diameter and have less amount of myoglobin and fewer mitochondria. The intermediate fibers present medium characteristics among the other types of fibers, but histologically they more closely resemble white muscles but are more abundant in red muscles. They have several mitochondria equivalent to that of the red fibers [3].

Physiological Changes Promoted by Constant Exercise

Regardless of the sports discipline practiced, constant exercise induces bodily changes, such as a faster energy metabolism (in training, the rate of oxidative phosphorylation in skeletal muscle is increased up to 100 times), a more solid and resistant bone mass, an increase in the size of mitochondria in muscle cells, cardiac hypertrophy, and regularly stable blood pressure. The resting heart rate of a high-performance athlete ranges between 30 and 40 beats/min; on the other hand, in an untrained person, the heart rate rhythm is 60 to 90 beats/min. In the lung, the amount of air that expires per minute increases due to the decreased oxygen content in the blood and the production of carbon dioxide (CO_2). The increase in oxygenation capacity leads to lung enlargement depending on the sports discipline practiced; muscle tone can be increased, inducing greater resistance to the accumulation of lactate in the blood and accelerating the breakdown of muscle lactate, generating resistance to muscle fatigue. Local changes in the muscles, such as the amount of lactate produced, an increase in muscle temperature and the concentration of CO_2 due to increased metabolism, facilitate the discharge of oxygen from the molecules of hemoglobin, which can also increase the number of red blood cells in the blood or the amount of blood plasma, which reduces blood viscosity. The mobilization of the different energy reserves to meet the demand of cardiac and skeletal muscles during exercise varies according to the type, duration, and intensity of the exercise performed, as well as the intensity and the physical condition of the individual [4].

Changes in Gene Expression during Exercise

Exercise induces changes in gene expression in different organs and systems, such as muscle, vasculature, and energy pathways. The different phases of training and post-exercise recovery stage also entail transcriptional changes that tend to make more efficient metabolic conditions. Some changes in gene expression during physical activity are summarized according to their metabolic involvement.

Genes that Regulate Muscle Size

Muscle growth is regulated by molecular mechanisms involving load-induced skeletal muscle and various genes. Among them α-actin is of the principal protein components of skeletal muscle; muscle overload leads to muscle hypertrophy due to increased synthesis of this myofibrillar protein. The α-actin promoter of skeletal muscle contains the serum response factor (SRF), a hypertrophy regulatory element that activates contractile protein genes in response to muscle overload. Serum response factor (SRF) is a factor of transcription that increases in hypertrophic muscles, and it undergoes homodimerization with the SRF, thereby enhancing α-actin mRNA transcription. This mechanism seems responsible for hypertrophy in the late phase of the exercise-induced changes induced by muscle overload. Other genes appear to be involved, and the phosphorylation of the p70S6K enzyme activates translation. This signaling pathway is activated only by the muscle-loading exercise. Likewise, muscle flexion with load increases the transcription of the liver and muscle isoforms of insulin-like growth factor (IGF), which induces muscle hypertrophy (Fig. **1**). The extracellular concentration of IGF-I regulates the activation of the enzyme phosphatidylinositol-3-kinase (PI3K), which regulates intracellular mediators such as Akt/PKB, with their kinase activity, activating genes involved in phenotypic changes attributed to weight-bearing muscular exercise.

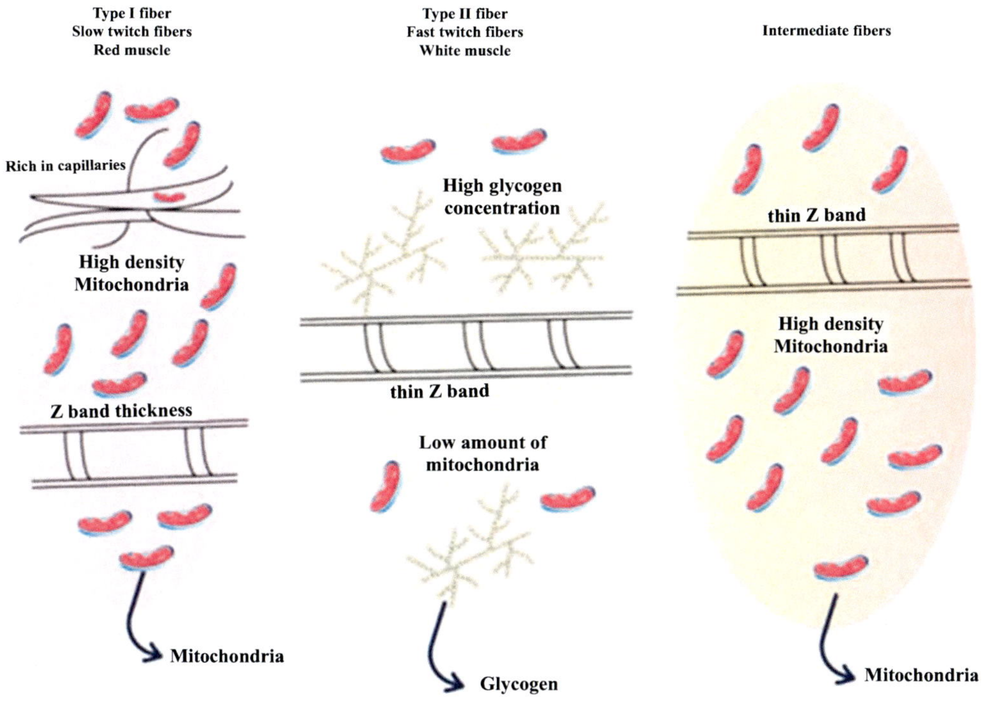

Fig. (1). Types of muscle fibers. Skeletal muscle is formed by combining three types of fibers called red, white, and intermediate, differentiated by the predominant metabolic pathway for obtaining energy. **A**) Type I fibers are abundant in red muscles and contain large amounts of myoglobin and numerous mitochondria. Fibers of this type are not very fatigable. **B**) Type II fibers in white muscles have less myoglobin and fewer mitochondria. They contain a large amount of stored glycogen compared to other muscle fibers. **C**) The intermediate fibers present average characteristics between the other types of fibers, but histologically they are more similar to type I and are more abundant in red muscle. They have several mitochondria equivalent to that of red fibers.

On the other hand, it is believed that the increase in calcium and IGF-I concentrations in load-bearing muscles increases calcineurin. Signaling the calcineurin pathway increases muscle mass, as this calmodulin-dependent phosphatase signals downstream and dephosphorylates the nuclear factor of activated T cells. Myostatin is also one of the genes modified during weight-bearing muscular exercise. This protein, known as the growth and differentiation factor, limits muscle growth and appears to retard the development of muscle stem cells.

Energetics Genes in Muscle

The duration and frequency of muscle contraction increase the number of mitochondria in the muscle. In this sense, cytochrome C is a marker of mitochondrial density in muscle; during exercise, the levels of mRNA, cytochrome C, and its protein increase and decrease in periods of inactivity. Adenosine monophosphate kinase (AMPK) activity also increases by two to three times during exercise. It is responsible for inducing some of the adaptations to resistance exercises, such as the increase in GLUT-4, hexokinase, and various enzymes involved in mitochondrial oxidation in muscle. In the sarcoplasm, increases in Ca2+ concentrations, such as those obtained during exercise, are associated with increased activity of the Ca2+ calmodulin-dependent kinase, which increases the expression of PGC-1 and cytochrome C. In the post-exercise recovery phase, genes involved in metabolism increase their transcription in skeletal muscle, such as pyruvate dehydrogenase kinase 4, heme oxygenase-1, lipoprotein lipase, and carnitine palmitoyl transferase I [5].

Cardiovascular Genes

Heart rate and blood flow increase during physical activity, requiring vascular dilation. To do this, the endocardium modulates vascular tone, synthesizing and releasing nitric oxide, which promotes vasodilation. The increase in nitric oxide-mediated by exercise is mainly due to gene and protein upregulation of endothelial nitric oxide synthase (eNOS). Another proposed molecular mechanism is that nitric oxide synthesized in aerobic exercise increases Endothelial superoxide dismutase (eSOD) mRNA levels in the adjacent smooth muscle cells, thereby preventing the degradation of nitric oxide mediated by O_2 radicals, and its half-life is prolonged. Likewise, constant exercise induces cardiac hypertrophy, the effect of which is to increase in cardiac reserve and contractility. Cardiac hypertrophy has the physiological advantage of improving cardiac function with reduced oxygen consumption, thus decreasing the submaximal and resting heart rate and increasing venous return and maximum cardiac flow. This ventricular growth induced by regular aerobic exercise brings blood flow from 75 ml/beat to 105 ml/beat, so it is possible to pump up to 30 L/min during physical activity. Calcineurin is partly responsible for this exercise-induced hypertrophy; athletes report this phosphatase activity in the myocardium up to 250% greater than in ordinary people. Similarly, IGF-I and norepinephrine could be involved in cardiac hypertrophy due to their effect on skeletal muscle.

Genes of Metabolism

AMPK is a sensor for ATP levels; low energy levels stimulate their activation, with which the signaling pathways are upregulated and will reset ATP stores.

AMPK increases in muscles with fast fibers in response to contraction under work; the downstream effects of this enzyme include stimulation of GLUT4 gene expression and hexokinase molecules that stimulate the oxidation of fatty acids and glycolysis, such as PFK2. It also negatively regulates the molecules that consume ATP, such as TORC2, glycogen synthetase, SREBP-1, and TSC2, which promotes the inhibition of gluconeogenesis and glycogen synthesis, lipids, and proteins.

Antioxidant Genes

In the acute phase of the exercise, vascular oxidative stress increases, with which reactive oxygen species (ROS) are generated, such as the superoxide radical (O_2^-) and peroxide of hydrogen (H_2O_2). Exposure of muscle cells skeletal to ROS stimulates glucose uptake *via* a phosphatidyl inositol-dependent mechanism 3-kinases (PI3K). O_2^- produced is the primary mediator for transmitting and activating hypoxia-induced factor 1 alpha (HIF-1a) signal in endothelial cells. This factor enters the cell nucleus, bound to hypoxia-induced factor 1 beta (HIF-1b), acts as a transcription factor for hypoxia-responsive genes (HRE) that code for a vascular endothelium growth factor (VEGF), erythropoietin (EPO), glucose transporter 1 (GLUT1), platelet-derived growth factor beta (PDGFb), the plasminogen activator inhibitor (PAI-I) and tumor growth factor alpha (TGF-alpha). These genes range from respiratory and transporter enzymes to hormones that regulate circulation and erythropoiesis, which are necessary for adaptation during exercise.

VO_2 max and Genetics

Fitness is measured by the maximum amount of oxygen an organism can carry in a minute and is defined by the maximum oxygen consumption (VO_2 max), measured in liters per minute, which correlates with the individual's weight and indicates the aerobic capacity of a person. VO_2 max improves with almost all types of training. Still, it depends on several physiological factors, such as the lung ventilation capacity, the lungs' diffusing capacity, the heart's size, blood's ability to carry oxygen, muscle utilization of oxygen, and the type of muscle fibers. VO_2 max is a capacity conditioned by genetics; various studies have shown that in people who respond to training, their VO_2 max increases by 60%. Other individuals, classified as non-responders, undergoing the same training do not significantly improve their aerobic capacity.

Angiotensin Converting Enzyme

The renin-angiotensin system plays a vital role in the homeostasis of the circulatory system produced by renal juxtaglomerular cells. Renin hydrolyzes angiotensinogen and generates angiotensin I, which will be converted into angiotensin II by the action of the angiotensin-converting enzyme (ACE). Angiotensin II is a potent vasoconstrictor that causes an increase in blood pressure. The renin-angiotensin system is also expressed in muscle and cardiac cells; ACE hydrolyzes and inactivates bradykinin, a vasodilator. The ACE gene is located on chromosome 17, site q23, and in its intron 16, there is an insertion gene variant (I) or deletion (D) of 287 bp. The D allele is associated with higher levels of circulatory disorders of ACE and, therefore, with increases in blood pressure and flow. Some studies have shown an increased frequency of the I allele in athletes who practice resistance sports and a higher frequency of the D allele in those engaged in sports that require great muscular force.

Some studies have shown that the skeletal muscle renin-angiotensin system can influence the overall body energy balance. When evaluating the efficiency of muscle contraction in healthy male individuals, after 11 weeks of training, it was observed that the energy efficiency increased by 8.6% in individuals with genotype II and only 0.39% in those with genotype DD. This effect is due to an association between the ACE gene DD phenotype and a higher average percentage of type I fibers (~50% vs. ~30% in genotype II).

BDKRB2 Gene

ACE is responsible for the creation of angiotensin II and the degradation of bradykinin. Therefore, bradykinin levels present an inverse relationship with the I/D polymorphisms of the ACE gene; higher concentration of ACE (DD allele) lower levels of bradykinin, and at lower levels of ACE (allele II), greater bradykinin concentration. Bradykinin levels influence glucose internalization and blood flow and prevent the growth of the heart's left ventricle *via* the activation of bradykinin-2 receptors (B2R). The −9/+9 variant of the B2R is located in exon 1 of the BDKRB2 gene, located on chromosome 14, site q32.1− q32.2, and refers to the absence (−9) of a 9bp segment associated with a high transcriptional activity that has consequently an increased expression of the receptor and, therefore, a heightened response to the agonist. Bradykinin modulates the left ventricle's hypertrophy response; allelic variations of the BDKRB2 gene are expected to alter the magnitude of this muscle growth.

APPLICATION OF MOLECULAR BIOLOGY IN SPORTS

The analysis of the genetic information of athletes or sports people seeks that high-performance athletes rely on molecular biology in the selection of their sports disciplines and helps coaches improve training programs according to the genetic characteristics of the athlete. Thus, the athlete will be focused on a job more specific and efficient according to their genetic predisposition. Research applying this technology to the sport since the 1990s is abundant and, through molecular biology, techniques such as polymerase chain reaction, are used for finding gene variants that will allow an athlete to excel in a certain sports discipline, with the purpose of channeling resources to their development, training, and sports preparation. It is considered that genetic factors determine up to 50% of muscular strength. However, the interpretation of this in terms of sports performance is questionable.

Sports performance is comprised, in addition to muscular strength, of several determining factors for success; however, it is recognized that people have genetic advantages for carrying out certain physical activities. The number of gene variations that potentially correlate with sports performance has increased, and 165 autosomal genes are currently known, five from the X chromosome and 17 from mitochondria. The genetic map of phenotypes associated with performance and conditioning is regularly published in Medicine and Science in Sports and Exercise. The last publication covered about 2000 genes, although it is considered only the tip of the iceberg compared to the role of the rest of the 23,000 coding genes in sports performance [6].

GLOSSARY

Molecular Biology: The study of the molecular basis of biological activity, such as the structure and function of genes and proteins.

Sports: Physical activities that involve competition and/or recreational activities.

Athletic Performance: The ability of an individual to perform physical activities related to sports or fitness, such as running, jumping, and strength training.

Genomics: The study of the complete set of genes, their structure, function, and interaction within an organism or a population.

Transcriptomics: The study of the transcriptome, the complete set of RNA transcripts produced by a cell or a group of cells.

Proteomics: The study of the proteome, the complete set of proteins expressed by an organism or a cell type.

Metabolomics: The study of the metabolome, the complete set of small molecules, such as metabolites and lipids, produced by an organism or a cell type.

Physical Traits: Observable physical characteristics of an individual, such as height, weight, muscle mass, and body fat percentage.

Strength: The ability of an individual to produce force against a resistance, such as lifting weights.

Endurance: The ability of an individual to sustain physical activity over an extended period of time, such as running a marathon.

Exercise: Physical activity that improves health and fitness, such as aerobic exercise, strength training, and flexibility exercises.

Health: The state of being free from illness or injury and having good physical and mental well-being.

Well-being: The state of being healthy, happy, and content with one's life.

REVIEW QUESTIONS

1. How can molecular biology be applied to the field of sports?
2. How does training affect gene expression in muscle cells?
3. Can molecular biology research help in the development of personalized training programs for athletes?

REFERENCES

[1] Booth FW, Chakravarthy MV, Spangenburg EE. Exercise and gene expression: Physiological regulation of the human genome through physical activity. J Physiol 2002; 543(2): 399-411. [http://dx.doi.org/10.1113/jphysiol.2002.019265] [PMID: 12205177]

[2] Higaki Y, Mikami T, Fujii N, *et al.* Oxidative stress stimulates skeletal muscle glucose uptake through a phosphatidylinositol 3-kinase-dependent pathway. Am J Physiol Endocrinol Metab 2008; 294(5): E889-97. [http://dx.doi.org/10.1152/ajpendo.00150.2007] [PMID: 18303121]

[3] Hohmann A, Lames M, Letzelter M. Introduction to the science of training. Barcelona: Paidotribe 2005.

[4] Kostek MC, Delmonico MJ, Reichel JB, *et al.* Muscle strength response to strength training is influenced by insulin-like growth factor 1 genotype in older adults. J Appl Physiol 2005; 98(6): 2147-54. [http://dx.doi.org/10.1152/japplphysiol.00817.2004] [PMID: 15894537]

[5] Rankinen T, Bray MS, Hagberg JM, *et al.* The human gene map for performance and health-related fitness phenotypes: The 2005 update. Med Sci Sports Exerc 2006; 38(11): 1863-88. [http://dx.doi.org/10.1249/01.mss.0000233789.01164.4f] [PMID: 17095919]

[6] Röckl KSC, Witczak CA, Goodyear LJ. Signaling mechanisms in skeletal muscle: Acute responses and chronic adaptations to exercise. IUBMB Life 2008; 60(3): 145-53. [http://dx.doi.org/10.1002/iub.21] [PMID: 18380005]

CHAPTER 12

Molecular Basis of Cancer

Maliha Sarfraz[1,*] and **Hayat Ullah**[2]

[1] *Department of Zoology, Wildlife and Fisheries, University of Agriculture Faisalabad, Sub-Campus, Toba Tek Singh, Pakistan*

[2] *Department of Chemistry, University of Okara, Okara, Punjab, 56300, Pakistan*

Abstract: Cancer is a disease that arises from the uncontrolled growth of cells due to genetic mutations and epigenetic changes. Molecular biology has provided valuable insights into cancer development and progression mechanisms. Cancer cells have alterations in the genes that regulate cell growth, division, and death, leading to the accumulation of mutations that confer a survival advantage. Oncogenes promote cell growth and division, while tumour suppressor genes inhibit cell proliferation and induce cell death. Alterations in these genes and changes in DNA methylation and histone modifications lead to the dysregulation of cell signalling pathways, which contribute to cancer development. In addition, the tumour microenvironment plays a critical role in cancer progression by providing growth factors, cytokines, and extracellular matrix components that promote tumour growth and invasion. Molecular biology techniques such as DNA sequencing, gene expression profiling, and epigenetic analysis have facilitated the identification of driver mutations and key molecular pathways involved in cancer development, leading to targeted therapies that exploit these vulnerabilities. Understanding the molecular basis of cancer can revolutionize cancer diagnosis, treatment, and prevention.

Keywords: Biomarkers, Cell growth, Cell signalling pathways, DNA methylation, Epigenetics, Genetic mutations, Gene expression profiling, Histone modifications, Oncogenes, Tumor suppressor genes, Targeted therapies.

INTRODUCTION

Cancer is characterized as a multifactorial disease that affects the average growth and proliferation of cells and produces alterations in the cell differentiation process, which determines the formation of a tumour in a specific tissue. Thus, the term cancer implies, in addition to uncontrolled cell growth and proliferation, a malignant transformation, *i.e.*, a loss of typical characteristics and functions of the cells in a tissue. These processes are influenced by genetic or epigenetic altera-

* **Corresponding author Maliha Sarfraz:** Department of Zoology, Wildlife and Fisheries, University of Agriculture Faisalabad, Sub-Campus, Toba Tek Singh, Pakistan; E-mail: maliha.sarfraz@uaf.edu.pk

Sanaullah Sajid, Sajjad ur Rahman, Shahid Mahmood, Shayan Bashir & Mudasser Habib (Eds.)

tions of numerous genes encoding proteins that regulate this process, contributing to a malignant phenotype. In addition, environmental factors are involved in tumorigenesis with a determining influence on some types of cancer. For a better understanding of cell cycle control, it can be schematized analogically as a biological clock, which operates from the nucleus of cells where various stimuli exist and/or converge that inhibit or induce cell cycle progression.

The cell cycle comprises eukaryotic cells' G1, S, G2, and M phases. Its regulation is a priority and is controlled by the participation of different proteins called cyclins and cyclin-dependent kinases (cyclin-dependent kinases, CDKs). Both act cooperatively, where CDKs phosphorylate certain cyclins at a specific time, allowing the cell to continue the cell cycle. In addition, it is important to consider that between the phases mentioned above, there are checkpoints that involve complex protein machinery that verifies whether the cell is given the right conditions or not to continue with the next cell cycle. These checkpoints control the progression between G1/S, G2/M, and M phases.

Cell cycle regulation involves cyclins and cyclin-dependent kinases as well as other proteins, which are the product of proto-oncogenes and tumour suppressor genes, which likewise participate in an organized manner in this control whose functions will be reviewed throughout the chapter. The complexity and importance of controlling the cell cycle, and their multiple participation ultimately complicate the re-establishment of an expected phenotype in malignant cells [1].

Proto-oncogenes, Oncogenes, and Tumor Suppressor Genes

Usually, cells respond to internal and external signals that stimulate cell proliferation, requiring regulation involving different cell cycle-stimulating proteins. Among the numerous genes involved in malignant cell transformation of cells include two large groups, classified as oncogenes and tumour suppressor genes. Both of them present a common denominator for their involvement in cell proliferation.

ONCOGENES

In 1910, Peyton Rous of the Rockefeller Institute and Nobel Prize winner experimented with chicken sarcoma. Injecting cell-free sarcoma homogenate into healthy chickens induced the same kind of tumour, raising the question of which agent, molecule, or factor was responsible for the tumorigenesis. The etiological agent of this condition was discovered in the following decades: a retrovirus called Rouss sarcoma virus (RSV). In 1970, one of the Rous sarcoma virus genes, the src gene, was described as having the ability to transform normal cells into malignant phenotypes. The study of this gene *via in situ* hybridization experiments performed by J. Michael Bishop and Harold E. Varmus at the University of California, San Francisco (1976) showed that normal chicken cells contained some elements of the src gene, which gave rise to the following controversy: how is it possible that genes found in the oncogenic virus (RSV) can also be found in normal eukaryotic cells? Subsequent studies showed that similar versions of oncogenic viral genes are found in normal cells and involved in cell proliferation. The src gene belongs to a tyrosine kinase family that regulates embryonic development and cell growth. These genes were named proto-oncogenes, and with this historical background, proto-oncogenes can be defined as genes that encode proteins that regulate regularly and physiologically to regulate the cascade of events that serve to maintain the control of cell cycle progression and the normal state of cell differentiation.

They are prefixed with the prefix "c" for "cellular" to distinguish them from the viral versions, which are prefixed with the prefix "v" (*e.g.*, c-src -cellular- and v-src-viral-). The altered versions of these proto-oncogenes by mutations give rise to oncogenes. As a result, the functions of proto-oncogene and oncogene gene products differ; in oncogenes, these products are oncoproteins. These proteins act pleiotropically by causing cellular and molecular changes, depending on the specific molecular level at which they act. Like oncogenes, they are determinants in the evolution of cancer. Tumour suppressor genes or anti-oncogenes play a crucial role in tumorigenesis since they restrict or suppress cell proliferation under certain conditions.

These genes are extremely important, as their inactivation or loss is necessary for malignant tumour formation, which confers their recessive nature. In 1986, the first retinoblastoma tumour suppressor gene (Rb) was cloned; two years later, the p53 gene emerged, a molecule that was known as "the molecule of the year" labelled by the scientific journal Science because of its importance in the physiological function. Advances in the study of oncogenes and tumour-suppressor genes have placed p53 at the centre of attention in studying carcinogenic processes. Thus, it is possible to speak of genes related to susceptibility to breast cancer, of a model that has established an association between histopathological lesions and the different mutations in proto-oncogenes and tumour suppressor genes, such as in colon cancer (multistep theory) and how the human papillomavirus is one of the main risk factors for the development of cervical cancer [2].

PROTO-ONCOGENES AND TUMOR SUPPRESSOR GENES OF BIOMEDICAL INTEREST

To understand the concept of the proto-oncogene, it is necessary to review the cascade of events leading to cell division. Growth factors are proteins that provide an external stimulus to the cell. These proteins, upon binding to their receptor on the cell membrane, internalize the signal by self-activating as an enzyme with tyrosine kinase or through second messengers such as G-proteins. It starts a phosphorylation cascade in cytoplasmic proteins, which eventually modifies transcriptional factors in the cytoplasm and causes them to migrate to the nucleus. After that, it binds to DNA and activates the expression of proteins that induce cell division (Fig. **1**).

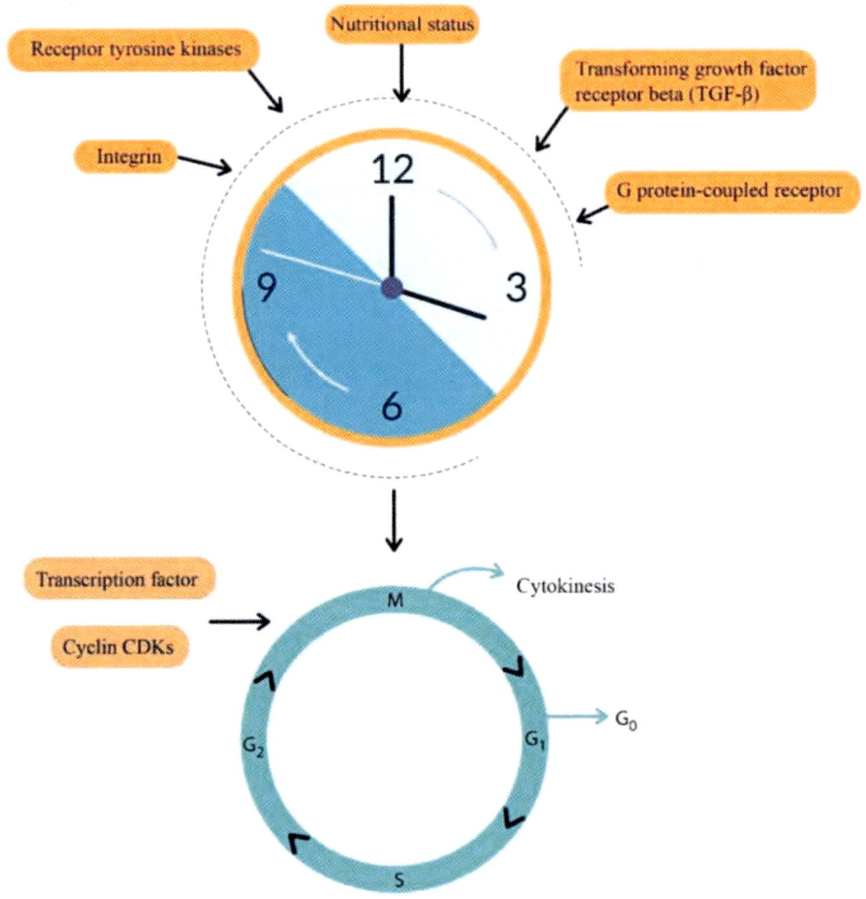

Fig. (1). Cell cycle stimuli. Analogous relationship between the cell cycle and a biological "clock": its progression depends on the participation of various stimuli.

In addition, inhibitory signals can slow cell division through similar mechanisms. Precisely, the balance between these two signalling cascades allows homeostasis of cell division in a normal cell, among other mechanisms. This cascade of signals is then coordinated by proteins from the extracellular space to the nucleus so that different cellular levels can be found where cell division is controlled, which are briefly described in Fig. (**2**).

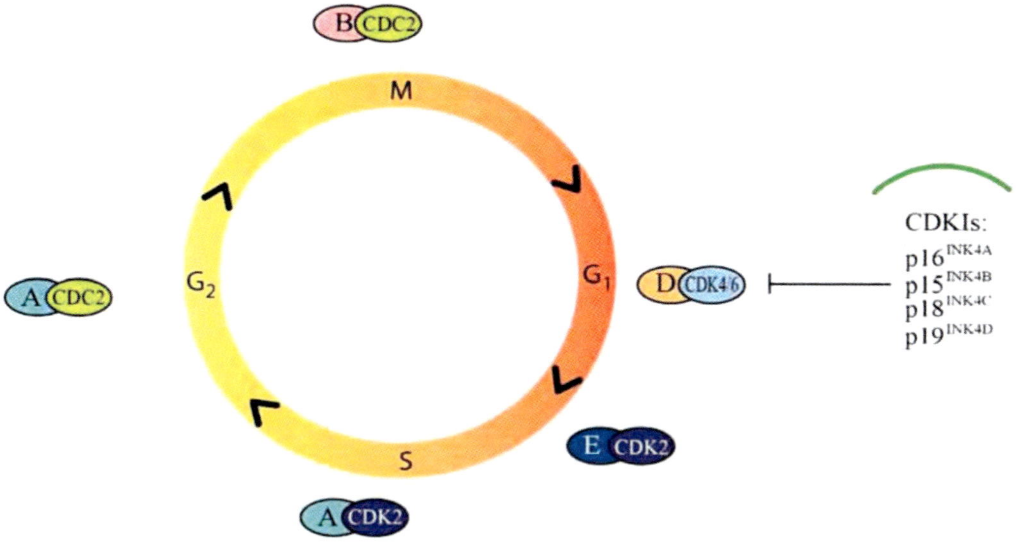

Fig. (2). Cell cycle: cyclins, CDK and CDKI. Different cyclin/CDK complexes exist throughout the transition from one cell cycle phase to another. CDKIs achieve inhibition of these complexes.

Growth Factors

Growth factors bind to extracellular domains of receptors embedded in the plasma membrane, where they trigger a transduction signal. The first oncogene described as a growth factor was v-sis, an oncogene from the simian sarcoma virus, similar to the β-chain of platelet-derived growth factor (PDGF).

Receptor with Tyrosine Kinase Activity

The epidermal growth factor receptor (epidermal growth factor, EGF) is a transmembrane protein with tyrosine kinase activity that is activated when the growth factor binds in the extracellular domain; consequently, the tyrosine kinase of the v-erb-B protein (the protein encoded by the avian erythroblastosis virus) is non-specifically activated without the need for stimulus; it thus loses its normal function and leads to oncogenesis. The epidermal growth factor receptor is found to be overexpressed or mutated in lung carcinoma, and it is the therapeutic target of a group of drugs called tyrosine kinase inhibitors (Fig. **3**).

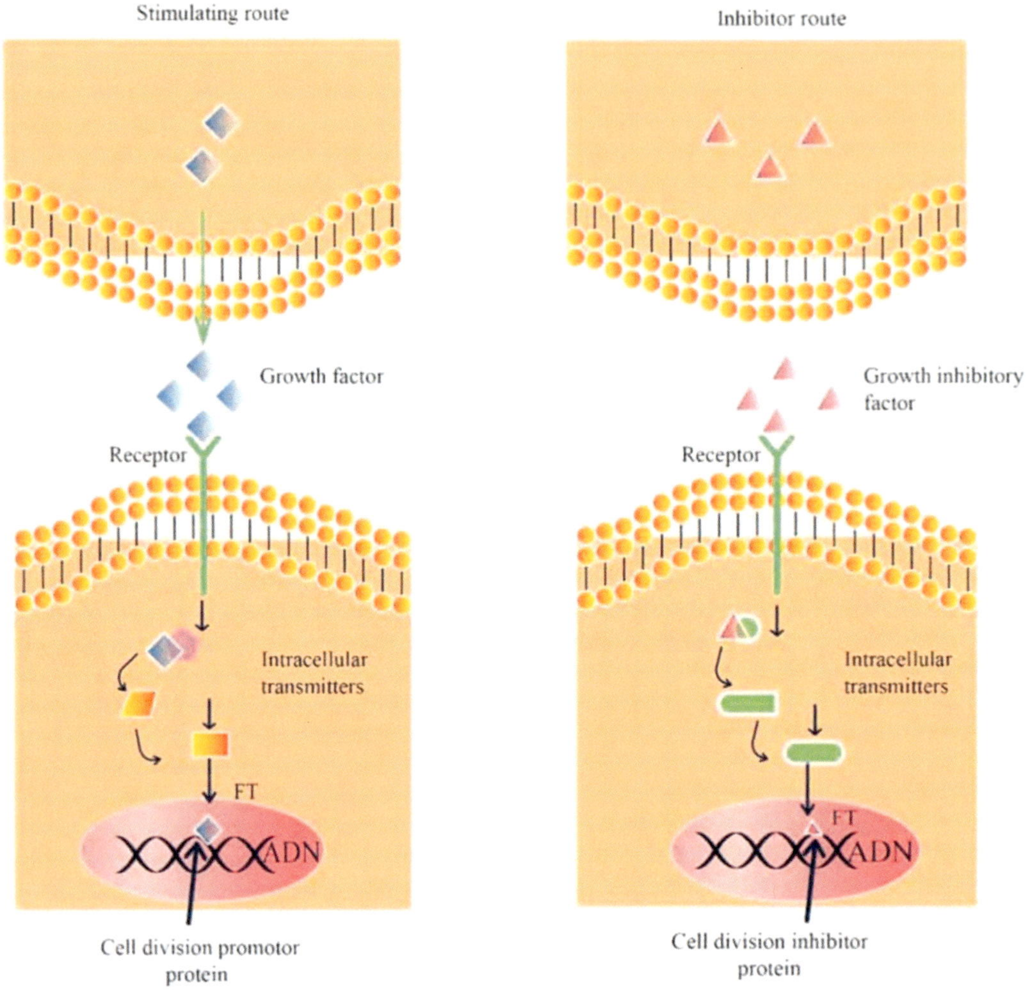

Fig. (3). Signal transduction. Factors that trigger signalling pathways that ultimately promote or inhibit cell division.

Membrane-Associated G Proteins

A typical example of such proto-oncogenes involved in signal transduction is the "ras" gene family, which includes three closely related proto-oncogenes, Harvey (ha)-ras, Kirsten (Ki)-ras, and N-ras. These encode for ~21 kDa proteins on the inner surface of the plasma membrane. Their function binds the GTP/GDP through GTPase activity, participating in signal transduction pathways that regulate cell growth.

Point mutations can activate the oncogenic potential of ras genes by altering the amino acid sequence of p21 proteins. *In vivo*, these mutations are found almost exclusively in codons 12, 13, or 61. However, point mutations have been described in other codons (59, 63, 116, and 119), whose transforming potential has been demonstrated by *in vitro* mutagenesis. Among the pathways that activate ras proteins include the MAP (mitogen-activated-protein) kinase signalling pathway. Notably, identifying ras mutations also contributes to the prognosis of different types of cancer, including colon and lung cancer, since the presence of ras mutations may contribute to tumours metastasizing earlier in their progression, which can serve as a prognostic factor for patients.

Cytoplasmic Kinases

Most kinase activity in mammals is carried out by phosphorylation of serine/threonine residues. These kinases are soluble cytoplasmic proteins and members of the signalling cascade that phosphorylate a protein by activating it; this, in turn, phosphorylates another, and so on, forming a cascade of phosphorylation that ultimately culminates in the activation of transcriptional factors acting in the nucleus; that is, they function as second messengers. The viral oncogene v-raf is a truncated version of the cellular protein. The amino-terminal regulatory region has a deletion that results in a constitutively active kinase.

Transcriptional Factors

Finally, the signalling cascade culminates in the activation of cell division-related genes through the activation/inactivation of transcriptional factors. The c-myc gene has been associated with various neoplasms in many species, and alteration in this gene is associated with multiple types of cancer, suggesting that it plays a crucial role in the biological processes of cell proliferation. In neoplastic processes, the c-myc gene acts by amplification, the protein is normal, but the expression is increased; in primary tumours, 8 to 10 copies of c-myc per Cell are reported. This phenomenon occurs even in the absence of growth factors. The presence of multiple copies of proto-oncogenes in tumour cells has been associated with a poor prognosis for patients.

Tumour Suppressor Genes

As described by Bert Vogelstein, a cancer researcher at Johns Hopkins University, a simple analogy that makes it easy to locate the two main types of genes involved in cancer is that an oncogene resembles the accelerator of a car. In contrast, a tumour suppressor gene can be considered the brake. In terms of the cell, when a cell speeds up, it divides, and when it slows down, the opposite occurs, which is why tumour suppressor genes are also known as anti-oncogenes. One difference between oncogenes and tumour suppressor genes is that the former only needs a mutated allele to activate their tumorigenic function, *i.e.*, they are the dominant character. In contrast, in tumour suppressor genes, both alleles of the gene must be altered to be non-functional, indicating that they are recessive. The mechanisms by which both alleles can be inactivated can be by the same type of mutation or by a different mutation.

Usually, two alleles of a gene are possessed without alterations; in some cases, a gene can be inherited with one of its alleles mutated; however, the other allele can replace the loss of function of the mutated allele. When this combination is lost by any of the mechanisms already mentioned, which may include random mutation, there is a loss of heterozygosity of the alleles; this implies changes in two mutated alleles, favoring the development of a tumour, initial condition for the development of cancer. The loss of heterozygosity is therefore related to the two-hit theory (established in 1971 by Alfred G. Knudson), which describes an increased likelihood of developing a specific type of cancer in individuals who inherit a mutated allele and whose cells are more susceptible to a somatic mutation in the normal allele. The effect of the two-hit theory on a cell is the initiation of an uncontrolled proliferation that gives rise to more cells with this characteristic. However, in cancer, the daughter cells with this mutation can have another type of mutation, either in the same gene or in other genes; this causes the proliferation to be even more rapid, or the cells do not differentiate according to their lineage. This gives rise to the clonal nature of cancer and highlights the biological complexity of this disease.

The best-characterized tumour suppressor genes in humans include the retinoblastoma (Rb) gene on chromosome 13 arm q14.1-q14.2 and the p53 gene on 17p13.1. There are other tumour suppressor genes of similar importance. Still, they are limited to specific cancer types, such as the APC and DCC genes located at 5q21-22 and 18q23.3, respectively, with clinical significance limited to colon cancer. The NF-1 and NF-2 genes, whose positions in the genome are 17q11.2 and 22q12.2, have been associated with neurofibromatosis.

Retinoblastoma

Retinoblastoma is a disease manifested by a malignant tumour in the retina since childhood. The karyotype of retinoblastoma patients showed a deletion of chromosome 13, where the Rb gene is located. It is worth mentioning that the malignant phenotype only develops when both Rb alleles are mutated. In the hereditary form of the disease, one of the parental alleles carries the deletion, and when the other (non-mutated) allele is altered, the tumour develops. It has been reported that there are up to 70% of cases of retinoblastoma with loss of heterozygosity in somatic retinal cells. There are also sporadic cases in which the parental origin allele is entirely normal. However, the tumour may be caused by somatic mutations in both alleles. Although a deletion causes the defect in most cases, other studies show that the defect may be in gene expression. The inactivation of the Rb gene has been reported in retinoblastoma, osteosarcoma, and small-cell lung cancer. In the case of alterations in gene expression, promoter methylation is found in the Rb in brain tumours. The Rb gene encodes for a nuclear phosphoprotein involved in cell cycle regulation through its cyclic binding to cyclin-CDK complexes, specifically cyclins D, E, and transcriptional factor E2F. During the end of the G1 phase, it phosphorylates and releases E2F; the E2F is phosphorylated and released, allowing transcription. During mitosis, this gene undergoes dephosphorylation, and this hypo-phosphorylated form inhibits cell proliferation (Fig. **4**).

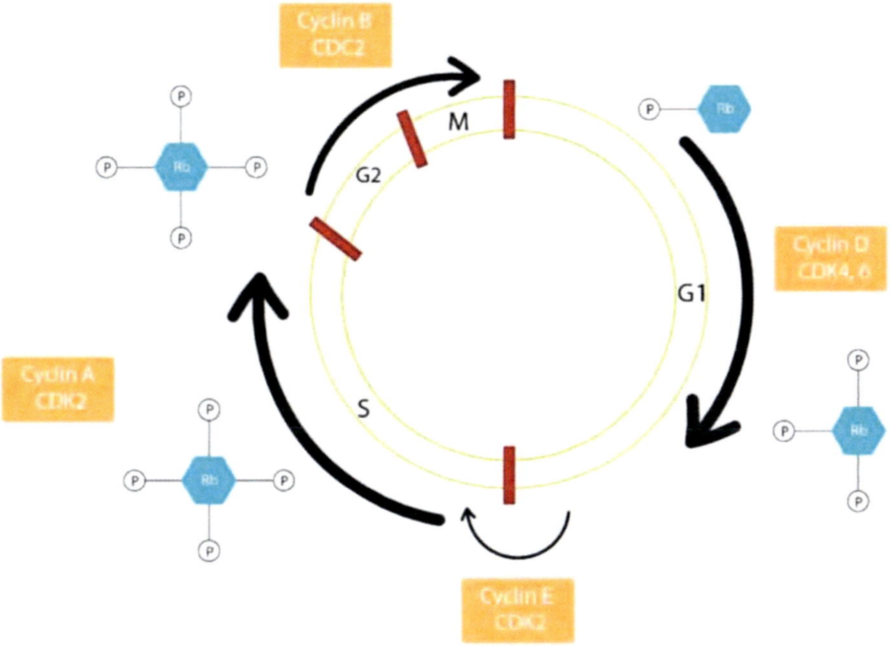

Fig. (4). Rb and cell cycle. The progression of the cell cycle depends on the phosphorylation state of Rb.

P53

The p53 gene encodes for a transcriptional factor that activates the transcription of specific genes and stops the cell cycle in response to DNA damage. This is achieved by the inhibition of CDK/cyclin complexes, as well as by the phosphorylation of Rb. If the DNA cannot be repaired, the cells increase their p53 levels and arrest in the G1 phase before entering the synthesis phase to activate the repair mechanisms. If the injury is not repairable, the cell is induced to apoptosis; cells with a repairable injury return to the S phase and cell division. The main objective is to prevent altered genetic information from being transmitted. Premalignant cells with a p53 mutant can communicate this alteration. As with Rb, cancer is more likely if mutations of the parental origin alleles are inherited. This inherited pathology is known as Li-Fraumeni syndrome and has an autosomal dominant inheritance pattern.

In 1987, Donna George reported that the MDM2 oncogene suppresses the ability of p53 to inhibit tumorigenesis. In this regard, MDM2 amplifications and p53 mutations have been reported in osteosarcomas. Because p53 is implicated in numerous cancer types, it has been proposed that its detection could be helpful as a marker in disease progression. The most frequently reported mutations in p53 are missense point mutations, which can occur throughout the entire reading frame, particularly in exons 5 to 8, constituting the protein's core region. However, each type of cancer has particular mutations that are more frequent at specific sites. For example, mutations have been detected in colon cancer at codon 165 and hepatocellular carcinoma at codon 249. The nonsense point mutations are usually associated with a prolongation of the half-life of the protein and, therefore, an increase in the basal level of the protein.

Finally, the loss of the activity of a tumour suppressor gene is always compensated by the activity of another gene that emerges as a safety mechanism in the face of emerging situations of the cells. However, these mechanisms are lost in tumorigenesis.

VIRUSES AND CANCER

Cervical cancer (CC) is a risk factor for human papillomavirus infection (HPV). Some studies report HPV-positive infection in more than 90% of patients with CC cases. It is worth mentioning that there are high-risk viral types (HPV 16 and 18, among others) and low-risk viral types (HPV 6 and 11, among others) to the development of CC. Because HPV infection occurs in the basal epithelial cells, it is maintained during the epithelial differentiation process to ensure the synthesis of viral proteins that will give rise to new viruses.

In response to infection, cells activate cellular defense mechanisms that involve p53 and Rb and direct the infected Cell to apoptosis. High-risk HPV viruses overexpress the viral oncoproteins E6 and E7, which inactivate tumour suppressor genes. E6 inactivates the p53 function, while E7 inactivates Rb. In both cases, there is a loss of its activity, which contributes to the tumorigenesis of CC, as changes in normal growth and differentiation of the cervical epithelium and genetic alterations accumulate in these cells. However, in addition to viral infection with a high-risk HPV genotype, it is important to consider the competence of the host immune system in eliminating the viral infection, as not all infections with high-risk HPV progress to CC. This highlights the importance of the humoral and cell-mediated immune response, like other factors recurrent infections, co-infections with different viral types and integration into cellular DNA.

Genetic Predisposition to Cancer

Cancer is passed down through families at a high rate. Infantile retinoblastoma, in which a mutated allele of the Rb tumour suppressor gene is inherited, is perhaps one of the most illustrative examples. More than 80% of cases of retinoblastoma develop between the ages of 3 and 4 years. However, some people who inherit a mutated Rb allele do not develop a tumour. This indicates a predisposition to cancer at certain stages of development, and such patients lose their susceptibility to the disease if they do not develop it in early childhood. Patients carrying an Rb mutation who did not have retinoblastoma in childhood develop non-ocular tumours later in life. Similarly, patients affected with Li-Fraumeni syndrome have a high risk of developing brain and breast tumours.

Another hereditary neoplastic syndrome related to defects in tumour-suppressor genes is adenomatous polyposis (FAP), whose inheritance pattern is autosomal dominant. This disease represents a model for study because of the correlation between the histopathologic lesions in the colonic epithelium and the genetic alterations in tumour suppressor genes (APC, MCC, DCC, p53) and oncogenes (ras). Patients with FAP inherit mutations in the APC gene, which stimulates the formation of polyps that tend to become malignant, as other genetic alterations in tumour suppressor genes are involved during malignant transformation, presenting what is known as loss of heterozygosity and clonal expansion. Clonal expansion refers to a cell's ability to initiate, promote, and convert to a malignant phenotype through the expansion of a mutated clone. The different types of cancer are complex diseases that are multigenic and, in some cases, cause a strong environmental influence [3].

Multiple Mutation Theory

As mentioned in the development of this chapter, tumorigeneses involve several events in which both oncogenes and tumour suppressor genes are involved. Tumour suppressor genes and cell differentiation have also been associated with numerous cell cycle control and differentiation genes. These genes have an impact on signaling pathways, immune response, apoptosis, and angiogenesis. The genetic alterations involved in malignant cell accumulation finally transform cells and give rise to a tumour. These multiple mutations, of a clonal nature, occur at different histological stages and have been correlated with evolution; in the case of colon cancer, one of the most representative models of this theory, one of the most representative models of this of this theory [4].

GLOSSARY

Absolute Risk: The difference between two risks is usually smaller than the relative.

Acute: Refers to symptoms that start and worsen quickly but do not last long.

Adjuvant Therapy: Treatment given after the primary treatment reduces the chance of cancer returning by destroying any remaining cancer cells. It usually refers to chemotherapy, radiation therapy, hormone therapy, and/or immunotherapy after surgery.

Benign: Refers to a tumour that is not cancerous. Cancer does not usually invade nearby tissue or spread to other body parts.

Biopsy: The removal of a small amount of tissue for examination under a microscope. Other tests suggest that cancer is present, but only a biopsy can make a definite diagnosis.

Cancer: A group of more than 100 diseases that can begin almost anywhere in the body, characterized by abnormal cell growth and the ability to invade nearby tissues.

Carcinoma: Cancer that starts in the skin or tissues that line the inside or cover the outside of internal organs.

Chemoprevention: Using natural, synthetic (made in a laboratory), or biological (from a living source) substances to reverse, slow down, or prevent cancer development.

Chemotherapy: The use of drugs to kill cancer cells.

Invasive Cancer: Cancer that has spread outside the layer of tissue in which it started and has the potential to grow into other tissues or parts of the body, also called infiltrating cancer.

Metastasis: The spread of cancer from the place where cancer began to another part of the body. Cancer cells can break away from the primary tumour and travel through the blood or the lymphatic system to the lymph nodes, brain, lungs, bones, liver, or other organs.

REVIEW QUESTIONS

1. Describe the importance of the S phase?
2. Name the genes whose expression is activated by p53?
3. Define loss of heterozygosity.
4. Investigate which susceptibility genes are involved in breast cancer and by what mechanism?

REFERENCES

[1] Hunter T. Cooperation between oncogenes. Cell 1991; 64(2): 249-70. [http://dx.doi.org/10.1016/0092-8674(91)90637-E] [PMID: 1988147]

[2] Jansson A, Gentile M, Sun XF. p53 Mutations are present in colorectal cancer with cytoplasmic p53 accumulation. Int J Cancer 2001; 92(3): 338-41. [http://dx.doi.org/10.1002/ijc.1189] [PMID: 11291068]

[3] Jones SN, Hancock AR, Vogel H, Donehower LA, Bradley A. Overexpression of Mdm2 in mice reveals a p53-independent role for Mdm2 in tumorigenesis. Proc Natl Acad Sci 1998; 95(26): 15608-12. [http://dx.doi.org/10.1073/pnas.95.26.15608] [PMID: 9861017]

[4] Weinberg RA. The biology of cancer. New York 2013. [http://dx.doi.org/10.1201/9780429258794]

CHAPTER 13

Molecular Mechanisms of Diabetes Mellitus

Itrat Fatima Toor[1,*], **Munazzah Sajid**[2] and **Haseeb Ullah Sajid**[3]

[1] *National Institute of Food Science and Technology, University of Agriculture, Faisalabad, Pakistan*

[2] *Department of Chemistry, University of Agriculture, Faisalabad, Pakistan*

[3] *Institute of Microbiology, University of Agriculture, Faisalabad, Pakistan*

Abstract: Diabetes Mellitus (DM) is a complex metabolic disorder characterized by high blood glucose levels resulting from defects in insulin secretion, insulin action, or both. The disease affects millions worldwide and is a leading cause of morbidity and mortality. Molecular mechanisms underlying the pathogenesis of diabetes mellitus are complex and involve multiple cellular and molecular processes. In this review, we discuss the current understanding of the molecular mechanisms involved in the development and progression of diabetes mellitus. Specifically, we focus on the role of pancreatic beta-cell dysfunction, insulin resistance, and abnormalities in glucose metabolism, lipids, and proteins. We also examine the contribution of genetic and environmental factors to developing diabetes mellitus. Additionally, we highlight the importance of targeting these molecular mechanisms for developing new and effective therapies for managing diabetes mellitus. A better understanding of the molecular mechanisms involved in diabetes mellitus can lead to more effective treatments and better disease management.

Keywords: Beta-cell dysfunction, Diabetes mellitus, Genetic factors, Glucose metabolism, Insulin resistance, Molecular mechanisms, Therapeutic targets.

INTRODUCTION

Type 1 diabetes is characterized by the destruction of (beta) cells of the islets of Langerhans due to an autoimmune mechanism or an unknown cause, while the phenomenon of insulin resistance characterizes type 2 diabetes. Both types of diabetes mellitus (DM) are endocrine, metabolic diseases characterized by chronic hyperglycemia and disturbances in carbohydrate, fat, and protein metabolism (DM type 2). In the first century, the term diabetes was used to refer to passing large amounts of urine (polyuria). The term was coined by the Greek philosopher Arateus; he called it diabetes mellitus or diabetes saccharin due to the sweet smell

* **Corresponding author Itrat Fatima Toor:** National Institute of Food Science and Technology, University of Agriculture, Faisalabad, Pakistan; E-mail: itratfatima71@gmail.com

Sanaullah Sajid, Sajjad ur Rahman, Shahid Mahmood, Shayan Bashir & Mudasser Habib (Eds.)

of urine; hence it is called in Latin: diabetes mellitus. In recent years, the terms insulin-dependent diabetes mellitus and non-insulin-dependent diabetes mellitus were previously applied to diabetes types 1 and 2, respectively, as both classes may require insulin administration for treatment [1].

EPIDEMIOLOGY

DM is currently recognized as a true epidemic, and it is estimated that in the world, about 200 million people have diabetes, it is predicted that this figure will rise to 300 million by 2025. The Latin American Diabetes Association (ALAD) indicates that the prevalence of diabetes mellitus in Latin America is from 4 to 16%, and an increase is expected in the next 25 years from 25 to 50%. In the United States, in young patients with obesity, the prevalence is 6.6%; 4% for individuals aged 20 to 60 years, 12% for those aged 60 to 64, and 17% for the elderly. An increase in the prevalence of both types of diabetes is expected worldwide. However, the incidence of DM2 may increase more rapidly due to obesity, reduced physical activity, aging of the population, and urbanization of societies. In Mexico, DM2 is the leading cause of premature death and permanent disability, with a prevalence of 10.9%; in addition, one in four people is unaware that they are suffering from the disease, and even with treatment, about 60% do not show optimal blood glucose control.

CLASSIFICATION OF DIABETES

The most commonly used etiological classification was recommended by the American Diabetes Association (ADA) in 1997 and by the Committee of Experts for the Classification and Diagnosis of Diabetes (World Health Organization, 1998) [2]. Both the organizations classified DM in:

DM type 1: It can be of 2 kinds: autoimmune (DM1a) and idiopathic (DM1b). It is also known as juvenile diabetes, as it often occurs during childhood, although it can also occur in adults.

DM type 2: It can vary between insulin-resistant, with a relative deficiency of insulin, to a defect, preferably secretor with or without RI corresponds to the majority (90%) of cases of diabetes. However, in most patients with DM2, multiple genetic and environmental factors contribute to the disease's origin, progression, and late complications.

MODY (Maturity-Onset Diabetes of the Young): It implies genetic defects in pancreatic B-cell function beginning in youth or maturity.

Gestational DM: This condition develops at any time during pregnancy in a woman without diabetes. It is the one that is presented in the course of pregnancy. It consists of the presence of elevated levels of glucose in the blood.

Childhood DM: It can be grouped into several main subtypes:

- Transient or permanent neonatal diabetes (TNDM and PNDM, respectively).
- DM1: DM1a, autoimmune, DM1b, idiopathic.
- DM2: This type of DM is reported in obese children.
- Mitochondrial diabetes (MDM).
- Cystic fibrosis-related diabetes (CFRD).
- Diabetes associated with syndromes such as Down and Turner.

Type 1 Diabetes Mellitus

This disease is characterized by the absolute deficiency of insulin, caused by an immune attack against the beta cells of the pancreas. The Islets of Langerhans are infiltrated with activated T-lymphocytes and cause insulitis. This autoimmune attack leads to the gradual destruction of pancreatic beta cells. Symptoms appear suddenly when 80 to 90% of the beta cells of the pancreas are damaged. The pancreas stop responding appropriately to the stimulation of glucose to secrete insulin, so it must be administered exogenously to restore metabolic control and prevent the ketoacidosis that can put the patient's life at risk. For this autoimmune attack to occur, it is necessary to present a stimulus in the environment (in some cases, it is due to a viral infection), in addition to a determinant genetic disorder that leads the immune system not to recognize the beta cells as their own. People with DM1 should administer exogenous insulin to live.

Type 2 Diabetes Mellitus

Hypertension with hyperinsulinism is already present in the prediabetic normoglycemic stage. It is a chronic disease with complications that involve a significant cause of mortality and are associated with the damage or failure of various organs. Genetic susceptibility factors are polygenic; however, environmental factors exacerbate these abnormalities. A reduced mass of beta cells in the pancreas and increased cellular apoptosis characterize individuals with DM2. Hyperglycemia in fasting results from hepatic resistance to insulin action. Postprandial hyperglycemia results from abnormal insulin secretion by the pancreas's beta cells by consuming food and the inadequate intake of glucose by the insulin-sensitive peripheral tissues, particularly in the skeletal muscle.

The lack of response of the receptor tissues to insulin maintains the stimulus of elevated blood glucose, which increases the production and release of insulin by pancreatic beta cells to compensate for and maintain normal blood glucose levels. This pathophysiological process of hyperglycemia and hyperinsulinemia and functional disruption of the pancreatic β-cell can last several years of the individual's life (up to 15 years), up to becoming unsustainable for beta cells. Hence, it presents glucose intolerance, with an exaggerated elevation of the postprandial glucose, IR, and, subsequently, the pathological process of the DM2. Metabolic alterations in DM2 are milder than those described for DM1, partly because the partial secretion of insulin, although insufficient, limits the presence of ketogenesis. The appearance of DM2 depends almost entirely on genetic and environmental factors, which are considered risk factors for this disease, as they are:

- Diabetic parents or siblings.
- Obesity and hypertension.
- Age over 45 years.
- Belonging to specific ethnic groups (African Americans, Native Americans, Asians, Pacific Islanders, and Hispanic Americans).
- Gestational diabetes or delivery of a baby with a body weight greater than 4 kg.
- Elevated levels of triglycerides in the blood.
- Elevated cholesterol levels in the blood.
- Lack of exercise and sedentary life.

The ADA recommends that adults over 45 get tested for DM2 at least every three years and, more often, in high-risk people. The diagnostic criterion of the ADA defines DM as fasting blood glucose ≥ 126 mg/dl or blood glucose > 200 mg/dl within 2 hours of a 75 g of the anhydrous glucose load.

Table 1. Comparison of the main clinical features between type 1 and 2 diabetes.

Clinical Features	Diabetes Type 1	Diabetes Type 2
Age of onset	Most often during childhood or puberty.	Usually after 35 years of age.
Development time	It develops quickly.	It develops gradually (chronic).
Nutritional status	Malnutrition often occurs.	There is almost always obesity.
Prevalence	900,000 = 10% of diagnosed diabetics.	10 million = 90% of diagnosed diabetics.
Genetic predisposition	Moderate	Very marked.
Defect or deficiency	Beta cells are destroyed, eliminating the production of insulin.	Insulin resistance, combined with the inability to produce adequate amounts of insulin.
Presence of ketosis	Frequent	Rare

(Table 1) cont.....

Clinical Features	Diabetes Type 1	Diabetes Type 2
Plasma insulin	Low or absent	High in the initial stages, low in the chronic phase
Complications	Ketoacidosis	Hyperosmolar coma
Treatment	The application of insulin will always be necessary	Oral hypoglycemic drugs, diet, exercise, and sometimes insulin.

Diagnostic Criteria for Type 2 Diabetes Mellitus

Regardless of the definition, various criteria for diagnosing DM are constantly revised by the ADA and the WHO. These criteria are summarized below [3].

Random Glucose Test

The existence of diabetes is suspected if the levels are higher than 200 mg/dl and are accompanied by increased thirst, urination, and fatigue (this test should be checked with other blood glucose levels in fasting).

Fasting Glucose Test for At Least 3 Hours

Diabetes is diagnosed if the result is more significant than 126 mg/dl twice. Levels between 100 and 126 mg/dl signify impaired fasting glucose or prediabetes. Such levels are considered risk factors for DM2 and its complications.

Glucose Tolerance Test

Glucose is measured at specified intervals. It starts with fasting glucose intake. Afterward, the patient takes moderate glucose in a sweet drink. It is diagnosed as diabetes if the glucose level is above 200 mg/dl after 2 hours (this test is most often used for DM2). Other tests include casual glucose, glucose fasting, postprandial glucose, tolerance curve to oral glucose, and glycosylated hemoglobin (GHB). Of all of these tests, the GHB or hemoglobin A1c (HbA1c) is an excellent measure of blood glucose control, especially over the long term. The importance of this test lies in the fact that HbA1c has a memory and indicates to the doctor the average glucose control over the last 2-3 months of testing.

Stages of Type 2 Diabetes Mellitus

Stage I. Normoglycemia compensated by hyperinsulinemia with different degrees of resistance.

Stage II. Baseline normoglycemia with hyperinsulinism is associated with decreased beta cell response pancreatic to glucose tolerance, generating postprandial hyperglycemia.

Stage III. Decreased insulin secretion with hyperglycemia in fasting [4].

Treatment of Type 2 Diabetes Mellitus

General therapeutic scheme: achieving good metabolic control, avoiding acute complications, and preventing chronic complications are the goals of the treatment to achieve self-monitoring and practical self-care, along with an appropriate emotional balance. The treatment is highly variable, depending on the patient's characteristics and the stage of the disease. Still, the general scheme should include physical exercise and a dietary regimen (without drugs) until these activities are combined with the help of different medicines, insulin administration, and combinations.

1. After positive diagnosis, changes are made in the diet and physical exercise routine.

2. If the above measures do not work (at least three months), drug treatment is started. In overweight people: Metformin is the drug of first choice. It inhibits hepatic glucose production and improves insulin sensitivity, which may be combined with sulfonylurea (SU), glitazone, secretagogue, or alpha-glucosidase inhibitors. People of average weight are recommended secretagogues to stimulate insulin secretion. They can use it alone or in combination with metformin, alpha-glucosidase inhibitors, or glitazone.

3. If drug treatment does not work, it should be given insulin alone or in combination with the medicines mentioned above.

Pathophysiology of Type 2 Diabetes Mellitus

The pathogenesis of DMT2 remains enigmatic. However, it becomes clear that lifestyle habits play a critical role, as evidenced by obesity as a risk factor. However, it is considered that genetic factors are even more critical than in DMT1. Studies on homozygous twins show that the incidence of DM2 in them has a concordance of 60 to 80%. The chance of acquiring DMT2 is 20-40% in patients with a family history of the condition and non-homozygous twins, compared to 5-7% in the general population. DMT2 is a heterogeneous syndrome with a common element, chronic hyperglycemia by a deficiency of insulin or insufficient effectiveness of its action. Glucoregulation is a complex physiological process in which failures at different levels ultimately lead to hyperglycemia.

Currently, defects have been identified in cells, tissues, or functions related to the expression of the disease or the presence of polymorphisms. Each of these modifications in the cells carries its own risk, which is influenced by environmental circumstances; the aggregate of these factors predisposes the disease's development to a greater or lesser amount. Obesity is a trigger factor for DMT2. About 80% of patients with DMT2 are obese; abdominal obesity (as opposed to obesity in deposits subcutaneously) is the one that presents a greater risk. The lipolytic action is opposed to the lipogenic effects of insulin and mobilizes fatty acids for fuel during fasting. The risk factors proposed to explain the increase in complications and incidence of cardiovascular disease in diabetes are hyperglycemia, dyslipidemia, accelerated formation of advanced glycation end products (AGE), such as HbA1c, the increase of oxidative stress, and genetic factors.

Insulin Function

Insulin is a pancreatic hormone released by beta cells of the islets of Langerhans in response to elevated levels of nutrients in the blood, which controls vital energy functions, such as glucose metabolism, lipids, and proteins. In addition, it promotes division and cell growth through its mitogenic effects. Insulin is a 5.8 kDa hormone, and it is composed of 51 amino acids arranged in two polypeptide chains, A and B, joined by two disulfide bridges. The A chain consists of 21 amino acids and the B chain consists of 30. It also contains an intramolecular disulfide bridge between the amino acid residues 7-7 and 20-19 of the A and B chains.

Insulin is released into the portal vein in response to increased blood glucose. Its primary function is to maintain the blood glucose concentration between 80 and 105 mg/dl, to favor the entry and storage of this nutrient in muscle and adipose tissues. The liver absorbs 50% insulin; the remainder stays in the bloodstream for 5 to 8 minutes in healthy adults. The rest of the glucose is stored in the liver as glycogen, inhibiting insulin production. The actions of insulin are mediated by cascades of intracellular signaling, in which the initial phosphorylation of the receptor on tyrosine residues (Tyr) leads to a series of phosphorylation/dephosphorylation events of kinases of Tyr and serine/threonine (Ser/Thr), which transmit the signal for the regulation of metabolic events within the cell.

The physiological function of insulin is to regulate the level of blood glucose. Regular meals can raise blood glucose concentration to 140 mg/dl. However, the blood insulin level can rise dramatically after a meal depending on the glycemic index. A compound meal will reduce glycemic response for most proteins, fat, and very few carbohydrates. In contrast, a meal rich in monosaccharides (simple carbohydrates, such as sugar) will cause a sizeable glycemic response. Beta cells of the pancreas can recognize the increase in glucose and release insulin in 30 seconds. Insulin binds to a protein transporter in the blood that converts the macronutrients and micronutrients into myocytes, hepatocytes, and adipocytes (Fig. **1**).

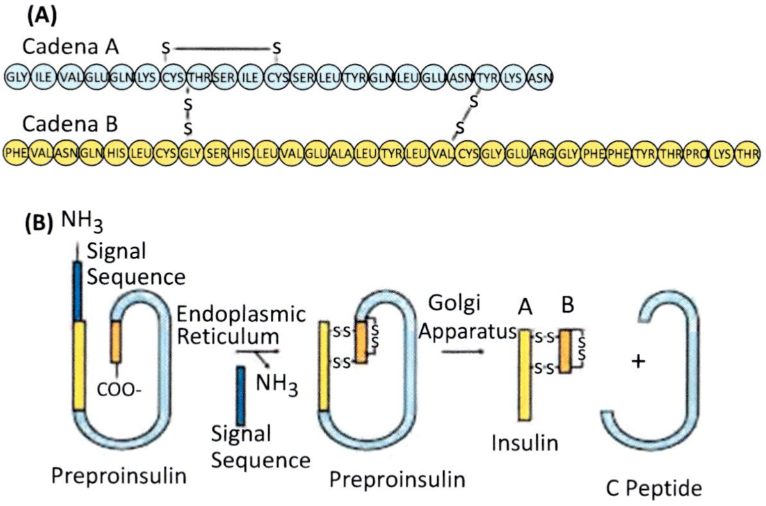

Fig. (1). Insulin. A) The structure of insulin. B) Formation of mature human insulin from preproinsulin.

Both obesity and lipoatrophy cause IR and a predisposition to DMT2, which shows that the adipose tissue is crucial for regulating the metabolism of carbohydrates beyond their ability to capture glucose. Myocytes have two glucose transporters: Glut 1 and Glut 4. It is considered that Glut 1 proteins are the primary transporters of glucose because its presence in the cell membrane does not vary. However, Glut 4 proteins are called inducible glucose transporters, as they move to the cell surface in response to insulin or muscle contraction. Physical exercise increases the number of Glut 4 proteins in the cell membrane and, consequently, insulin sensitivity, so much so that insulin manages to restore glucose level in the blood in 2 hours by introducing glucose into those cells.

Although insulin does not stimulate glucose uptake in the liver, it blocks glycogenolysis and gluconeogenesis and stimulates glycogen synthesis; in this way, it regulates fasting glucose levels. The action of insulin on tissues generally not considered sensitive to the insulin, including the brain and pancreatic beta cells, is also essential in glucose homeostasis.

Insulin Resistance

Insulin resistance can be defined as cells' inability to respond to insulin due to defects in its signaling. This IR is one of the main features in the pathological manifestations associated with DMT2. Multiple abnormalities have been identified in the insulin signaling reactions that contribute to this resistance:

- Reduction of the expression level of insulin receptors.
- Reduction in the level of tyrosine kinase intrinsic activity, resulting in phosphorylation of the inadequate substrate of the insulin receptor. Among the potential causes for the dysfunction of the beta cells of the pancreas, some areas follows:
- Reversible metabolic abnormalities such as glucotoxicity and lipotoxicity.
- Hormonal changes such as inadequate action of the incretin and the increased secretion of glucagon (the incretin hormones are produced in the gastrointestinal tract and are released when nutrients enter the intestine and, upon its release, the secretion of insulin is stimulated; the primary regulatory mechanism for the glucagon secretion is the level of glucose in the blood).
- Genetic abnormalities of beta cell proteins of the pancreas, such as glucokinase, the complex formed by the sulfonylurea receptor and the potassium channel.
- Reduction in the mass of beta cells of the pancreas mainly caused by apoptosis.

Insulin resistance increases with increasing weight but is reduced with the decrease. This suggests that the accumulation of fat is essential in the development of insulin resistance. The adipose tissue is an organ that stores energy and is a secretory organ. The regulatory substances that produce adipocytes include leptin, resistin, and adiponectin, which contribute to the development of resistance to insulin. The high levels of fatty acids in obesity are also involved in developing insulin resistance. The harmful effect of fatty acids on insulin signaling is carried out by activating different serine kinase proteins (JNK, IKK, PKC θ, and PKC ζ), which phosphorylate to the insulin receptor substrate (IRS-1). Therefore, the serine phosphorylation of IRS-1 plays a crucial role in insulin resistance induced by fatty acids. Preventing this phosphorylation protects the muscle against insulin resistance. Adipose cells express and release the growth factor of tumor necrosis alpha (TNFa), an element whose participation is recognized in the induction of IR. This molecule exerts multiple effects on adipocyte function, such as inhibiting lipogenesis and increasing lipolysis. The TNFa interferes with insulin signaling and causes resistance to insulin in obesity.

Insulin Signaling

Insulin exerts various pleiotropic effects on cells: it increases cell replication and survival, decreases apoptosis, participates in the cell cycle, and regulates the homeostasis of energy metabolism. Insulin resistance may be due to multiple defects in the transduction of the signals, such as the faulty activation of the insulin receptor and the decreased activation of the kinase of phosphatidylinositol triphosphate (PI3K) stimulated by insulin. Several molecules are being investigated as a strategy for intervening in the transduction of insulin-mediated signals. New approaches aim to inhibit enzymatic pathways that deactivate the insulin receptor or their underlying intracytosolic effectors, such as proteins IRS. In this way, it has been possible to identify proteins of specific tyrosine phosphatases, among others, as therapeutic goals [5] (Fig. **2**).

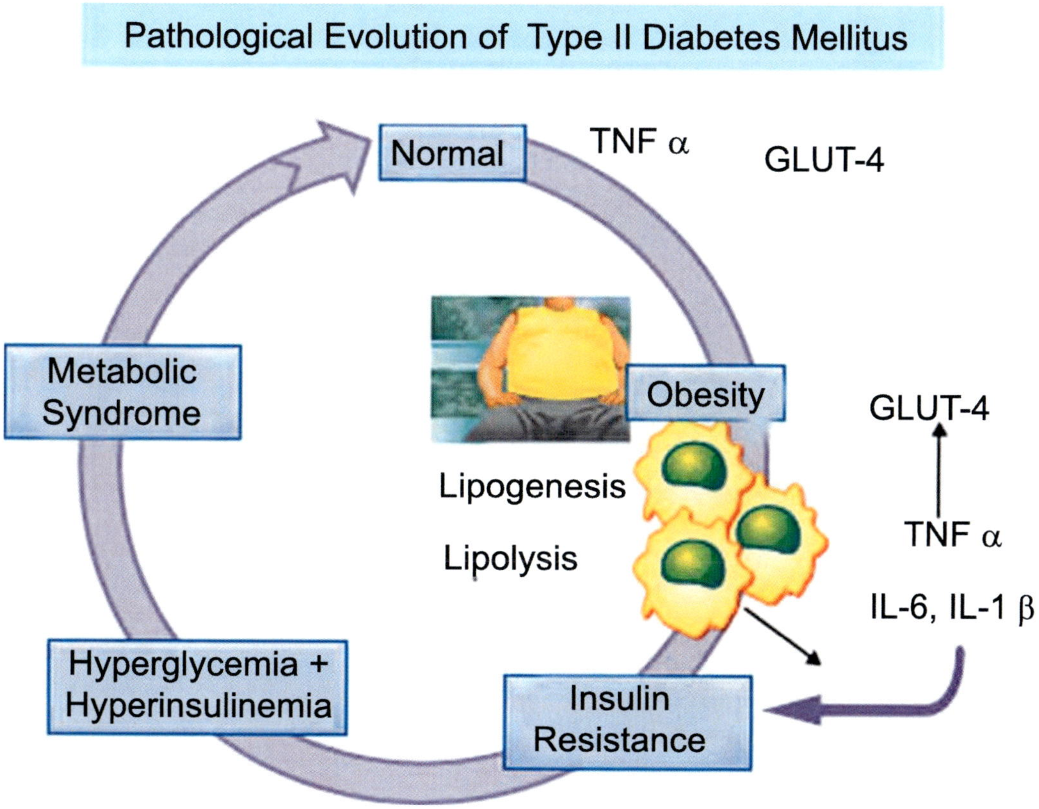

Fig. (2). Pathological evolution of type 2 diabetes mellitus. TNFα interferes with insulin signaling and causes insulin resistance in individuals with obesity. In obese humans, TNFα activity is increased. It can induce insulin resistance through a defect in the ability of the insulin receptor to phosphorylate or by the decrease in the gene expression of transporters of glucose Glut 4. TNFα is a stimulant of lipolysis, while it inhibits lipogenesis.

Insulin Receptor

Once insulin in the blood reaches the two organs' main target: the cells of the adipose tissue and those of the muscles, the latter is the prominent consumer of glucose. Muscle cells have specific receptors for insulin on their surface. The actions of insulin begin with its coupling to the insulin receptor (InsR), a transmembrane heterotetrametric protein composed of two extracellular subunits, alpha, and two transmembrane beta subunits, with tyrosine kinase activity. Insulin binds to two asymmetric sites of the extracellular alpha subunits, with which conformational changes are induced, resulting in the autophosphorylation of the inserted beta subunits across the membrane and causing the activation of the intrinsic tyrosine kinase receptor (phosphorylating enzyme). At this time, the intracellular portion of the receptor acquires the ability to kinase and makes a molecule of phosphate (H_3PO_3) combined with the amino acid tyrosine in each subunit (Fig. **3**).

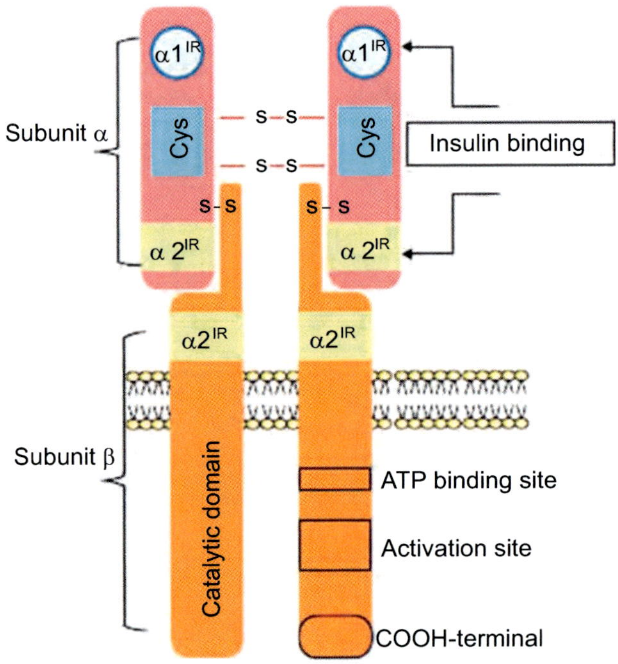

Fig. (3). Structure of the insulin receptor: IR is a heterotetramer with two extracellular α-subunits linked to two β-subunits by disulfide bridges. The α subunits contain the A1IR and a2IR insulin binding regions and a rich cysteine region (Cys). The α subunit contains an extracellular, one transmembrane, and one intracellular portion. In his intracellular part, a tyrosine catalytic kinase domain is located with an ATP-binding site and phosphorylation sites in the juxta membrane regions, activation site, and carboxyl-terminal.

The beta subunit has three autophosphorylation sites: one near the membrane, the other inside the regulating loop, and the third at the carboxyl-terminal end. Autophosphorylation starts a cascade of cell signaling responses, including the phosphorylation of the family of IRS proteins, of which at least four have been identified with similar chemical structures. Still, they are distributed differently in body tissues. The insulin-like effect is given by activating the phosphorylated intracellular chain by exchanging phosphors between various molecules within the cell. Due to insulin binding to the alpha subunits, the beta subunit, with protein kinase properties, auto phosphorylates on tyrosine, serine, and threonine residues. Subsequently, a cascade of phosphorylation reactions is initiated, and dephosphorylation of many intracellular substrates' intermediaries of many metabolic actions also occurs. This way, insulin achieves its effects through cellular events involving diverse reactions. Phosphorylated IRS-1, in turn, binds to various proteins called SH2 proteins, which contain SH2 domains, homology to domain 2 of the Src protein. Once phosphorylated, the InsR attracts adaptor proteins containing phosphotyrosine-binding domains (PTB), such as IRS and Src homologous proteins. Adapters protein that binds to the phosphorylated receptor can activate the mitogenesis-activating protein kinase pathway (MAPK) or the PI3K signaling pathway.

In the phosphatidylinositol-3-kinase signaling pathway (PI3K), the substrate proteins of the InsR, such as IRS1 and IRS2, bind to the phosphorylated tyrosine of the InsR. This turns them on so that they can attach to the SH2 domain of the PI3K of class IA, which binds to phosphotyrosine residues on activated receptors and adaptor proteins. The p85 subunit of PI3K is released when the p110 subunit of PI3K binds to phosphotyrosine on substrate proteins of the InsR. The p110 subunit of PI3K stops the p85 subunit from doing its job. Thus, the p85-p110 heterodimer can be attached to its phosphatidylinositol-4,5-bisphosphate (PIP2) substrate in the plasma membrane and produce phosphatidylinositol-3,4,- -triphosphate (PIP3).

The synthesis of PIP3 leads to the recruitment of serine kinases/threonine, such as Akt (also called protein kinase B or PKB), which can bind to phosphatidyl lipids inositol in cell membranes. Akt is a member of a family of proteins (there are three genes: Akt1, Akt2, and Akt3) and is phosphorylated by PDK1 (mTOR/Rictor protein complex), at the threonine residue 308 and serine 473, which activate it. Akt acts on several proteins by regulating protein synthesis, growth and cellular metabolism, neuronal plasticity, and apoptosis. AKT also can work on glycogen synthetase kinase 3 (GSK3), holding the cell cycle progression.

Insulin Receptor Substrates

At least nine intracellular insulin/IGF-1 receptor kinase substrates have been identified. Four of them belong to the family of receptor IRS proteins of insulin. Others include the proteins Gab-1, p60dok, Cbl, APS, and isoforms of Shc. The phosphorylated tyrosine in these substrates acts as docking zones for proteins that contain SH2 domains.

Inhibition of Insulin Receptor Signaling

In addition to tyrosine phosphorylation, both the receptors for insulin, like IRS proteins, are susceptible to phosphorylation on their serine residues, which can attenuate or abolish the intracytosolic signaling pathway of insulin, by decreasing or interrupting tyrosine phosphorylation mediated by insulin and also promoting interactions within 14-3-3 proteins. The 14-3-3 family of proteins is involved in signal transduction by binding to phosphoserine-containing proteins.

The action of insulin is also attenuated by proteins tyrosine phosphatases (PTPases), which catalyze a dephosphorylation structure of the receptor and its substrates. The 14-3-3 proteins act as modulators of phosphorylation; this means that after a protein is phosphorylated, a binding site for 14-3-3 is generated, and this interaction modifies the activation state of the protein. This phosphorylation negatively feedbacks the insulin signaling and serves as the molecular mechanism bridge for cross-communication of other intracytosolic pathways involved in insulin resistance. Several kinases have been implicated in this process, including PI-3K, Akt, the glycogen synthetase kinase GSK-3 and the rapamycin receptor in mammals (mTOR). On the other hand, the attenuation of signaling of adipose tissue-induced insulin could be triggered through different molecular and genomic pathways that cause the sequential activation of protein kinase-C (PKC) and the inhibitor of nuclear factor kappa-B kinase.

Stimulated Phosphorylation Cascades by Insulin

As other growth factors, insulin stimulates the MAPK, which enables the kinase to be regulated by extracellular signals (ERK). This pathway involves tyrosine phosphorylation in the IRS or Shc proteins, which interact with the Grb2 adapter protein, recruiting the son-of-sevenless exchange (SOS) protein towards the plasma membrane to activate the Ras protein. The activation of Ras requires stimulation of tyrosine SHP2 phosphatase through its interaction with substrates of the recipient as Gab-1 or IRS1/2. Once activated, the Ras protein operates as a molecular switch, stimulating a cascade of serine kinases through the sequential activation of Raf, MEK, and ERK proteins. The activation of ERK translocates it to the nucleus's interior, where it catalyzes the phosphorylation of transcription factors, which give rise to cell proliferation and differentiation.

Diffusion Transporters for Hexoses

The facilitated diffusion transporters for hexoses (Glut) are a large family of proteins whose members have common but extremely conserved characteristics that determine secondary and tertiary structures (domains or motifs). These domains give them specificity to transporters by one or more carbohydrates, determine their tissue distribution, the location of the cell, and the regulation of its activity by hormones. In the last four years, new genes encoding TDFH-like transporter proteins for glucose belong to the SLC2A family (Solute Carrier 2A). Based on the homology of the primary sequence, the Glut family can be divided into three subfamilies:

a. Class I family is formed by the Glut 1, 2, 3, and 4.
b. Class II family: made up of Glut 5, 7, 9, and 11.
c. Class III family: formed by Glut 6, 8, 10, 12, the Glut 13, or myoinositol transporter (HMIT-1), and the Glut 14.

Glut 1 (SLC2A1): Transport with High Affinity

It is found in tissues that use glucose as a primary fuel. Glut 1 is also known as the erythrocyte/brain glucose transporter. It is widely distributed in humans and is expressed in numerous fetal and adult tissues, such as erythrocytes, endothelial cells, nerve cells, placenta, white blood cells, retinal cells, kidney (mesangium), adipose tissue, *etc*. Glut 1 has a high affinity for glucose so that it can be transported inside the cells at any concentration. It is considered a basal transporter, essential for the nervous tissue and the erythrocyte.

Glut 2 (SLC2A2): Glucose Sensing Function

Glut 2 is a glucose transporter with low affinity and is expressed in the adult human liver, the kidney, the beta cells of the islets of Langerhans, and the basolateral membrane of the intestinal epithelial cells. The property of the glucose sensor is attributed to a low glucose concentration in plasma. It cannot transport glucose into the beta cell, and insulin secretion is deficient. Instead, after a meal, when the plasma concentration of glucose is increased, the ATP generated by the metabolism can stimulate insulin release. Glut 2 is bidirectional, transporting glucose from the blood to the tissue or from the tissue to the blood, which is very useful for the liver and kidney.

Glut 3 (SLC2A3): High Affinity for Glucose

Glut 3 is a high-affinity glucose transporter first characterized in the brain. It is assembled with Glut-1 in the nerve tissue and the placenta and participates in embryonic development.

Glut 4 (SLC2A4): Glut with high mobility

Glut 4 is a high-affinity transporter for glucose which is expressed mainly in the tissues of striated muscle, cardiac muscle, and adipocytes. Glut 4 represents an exceptionally reliable mechanism of regulation of glucose metabolism, which only allows the entry of glucose to muscle tissues when it is sufficiently elevated to stimulate the secretion of insulin that will favor the entry of excess glucose into the muscle. The mechanism of Glut 4 is through the binding of vesicles with the fixed plasma membrane. The translocation of Glut 4 to the membrane involves the activation of the enzyme phosphatidylinositol-3-kinase (PI-3K) through the phosphorylated IRS-1, which forms a complex with the said enzyme, which produces an increase in its activity about 20 times. This complicated system requires the participation of several vesicle and membrane proteins, called Snare proteins.

Glut 5 (SLC2A5): specific transporter for fructose

It is expressed mainly in the cells of the brush border of the small intestine, where it mediates the passage of fructose from the lumen to the intestinal epithelial cell. Low levels of this transporter are also found in erythrocytes, kidneys, spermatozoa, the skeletal muscle, and adipose tissue of humans and rats.

Glut 6 (SLC2A6): Low-Affinity Glut 6, As in The Case Of Glut 23

The most similar Glut 6 protein is the Glut 8 transporter (44.8% homology).

Glut 7 (SLC2A7): A Failed Number

After analyzing the Glut 7 sequences of rats, it was concluded that neither rat nor human hepatocytes contained the mRNA equivalent to the one cloned in the experiments initially, so it is inferred that the initial finding of Glut 7 was an artifact of the cloning techniques used at the time. Later a gene was detected with similar characteristics to that of the Gluts adjacent to the Glut 5 gene, with remarkable similarity (58% homology). This gene has been tentatively named SLC2A7, although the expression pattern of the respective gene and the specificity of its transcription by substrates have yet to be discovered.

Glut 8 (SLC2A8): The Race Starts to Discover New Glut

Finding deficient mRNA levels from the isoforms, in addition to these transporters, raises the existence of new carriers. Experiments on mice that do not express Glut 4 (Glut 4 knockout mice) showed that the glucose transport capacity in the muscle is almost normal, with no compensatory increase observed in the expression of Glut 1 or 3. This approach led to identifying and characterizing Glut 8 and Glut 9 (currently reclassified as Glut 6).

Glut 9 (SLC2A9): The Actual Glut 9

Glut 9 is a class II transporter with a 55% homology with Glut 5. It is expressed as necessary for the kidney and liver and at low levels in the small intestine, placenta, lung, and leukocytes.

Glut 10 (SLC2A10): Makes Coupling with The Glut 2

The transporter gene is located in a region strongly associated with possible diabetogenic. The location of the gene and its functional studies suggest that Glut 10 can carry out metabolic functions of great importance and can be a crucial element in developing DM2. This carrier is found in higher concentrations in the liver (adult and fetal), the pancreas, heart muscle, lung, brain (adult and fetal), skeletal muscle, placenta, and kidney.

Glut 11 (SLC2A11): A Transporter More than Fructose Only

Only products of the human Glut 11 gene have been obtained in the skeletal and cardiac muscle; three types of variants: the first is caused by the existence of three different initiation exons. Unlike Glut 4, the glucose transport activity of Glut 11 is significantly inhibited by fructose, a transporter of fructose with a low affinity for glucose.

Glut 12 (SLC2A12): Younger Brother from the Glut 4

The cDNA of this Glut encodes for a protein of 617 amino acids that possess the essential characteristics of Glut. The degree of homology of this transporter with Glut4 is 29%, and with Glut 10, it is 40%. Glut 12 features ample extracellular space between the transmembrane helices-α 9 and 10. The immunoblot has shown the expression of Glut 12 in skeletal muscle, adipose tissue, and the small intestine. This fact has raised the hypothesis that this transporter represents the second elusive insulin-sensitive transport system found in muscle cells, as its mRNA has been found in the muscle and the prostate.

Glut 13 (SLC2A13): Transporter of Myoinositol Within the Classification Of The Glu

Inositol and its phosphorylated derivative compounds (phosphoinositide) have essential functions as osmolytes and as second messengers in regulating exo and endocytosis of vesicles. Glut 13, or H+/inositol transporter, encodes for 629 amino acid membrane transporter protein with an analogy of 35% with Glut 6, which is strongly expressed in glial cells and some neurons, with the ability to transport myoinositol and glucose when found at a high concentration. The expression of this transporter in Xenopus laevis oocytes has shown that the transport activity is almost exclusively for myoinositol and some of its isomers, with a length of 100 mM. Its preferential expression in the central nervous system suggests its prominent role in regulating these metabolites in the brain.

Glut 14 (SLC2A14): A Distant Coding

This Glut is encoded about 10 Mb upstream of the Glut 3 gene, which shares a strong resemblance. Until now, it had been believed that Glut 14 was a pseudogene (like Glut 6 in the beginning) due to the duplication of the Glut 3 gene. The Glut 14 gene possesses two forms: a short one, consisting of a transcript of 497 amino acids, similar to Glut 3 by 94.5%. The second way, called the long form, encodes for a protein of 520 amino acids that differ from the previous one at the amino-terminal end. However, in contrast to Glut three, this transporter is mainly expressed in the testicles, where its mRNA is found at a concentration four times greater than Glut 3.

Susceptibility Genes to Diabetes Mellitus Type 1 and 2

The application of molecular biology techniques has advanced the search for genetic markers associated with various diseases for their timely diagnosis and treatment. In contrast to DM1, which has been found to have a close relationship with the genes of the HLA system, in the DM2, it has not been possible to establish a single gene involved with this disease. However, several genes related to this condition have been described, including genes that encode for the proteins involved in the insulin signaling pathway, glucose transport, glycogen synthesis, the synthesis and absorption of fatty acids, and the differentiation of adipocytes. Less common forms of DMT2 are MODY and autosomal dominant inheritance (Mendelian forms). This subtype of diabetes, usually early onset, like diabetes of complex heredity, is genetically heterogeneous. To date, six genes have been described as causes of different MODY: the gene that encodes for the glucokinase enzyme, an essential regulator of secretion of insulin by the pancreas, and five more genes that encode for transcriptional factors involved in the expression of the insulin gene and other pancreatic genes, such as the hepatocytes nuclear factor (HNF) 4 alpha, (HNF-4α), HNF-1α, HNF-1β, IPF-1, and beta2.

Diabetes and Physical Activity: Effect on Glut 4, Glycogenolysis, and Insulin Resistance

The beneficial effects of physical activity on treating DM2 are recognized. Several types of research have shown that aerobic exercise is a form of very convenient training. However, nowadays, in patients with DM2, the exercise of overload is an essential component of physical program conditioning. Myocytes have two glucose transporters, Glut 1 and Glut 4. Glut 1 proteins are considered the basal level transporter of glucose because their presence in the cell membrane does not change. In other words, muscle cells have a certain number of proteins, Glut 1, on the cell membrane to transport glucose. However, Glut 4 transporters move to the cell's surface in response to insulin or muscle contraction induced by physical activity. Thus, physical exercise increases the number of Glut 4 proteins in the cell membrane and insulin sensitivity, as it is translocated into the muscle cell membrane without insulin intervention. Simple walking allows the entry of 100 g of glucose/hour to the muscles, while intense exercise allows the intake of 200 g of glucose/hour.

AMPK AND MUSCLE GLUCOSE CONSUMPTION

AMP-activated protein kinase (AMPK) is an enzyme activated when there is a change in the cell's deposit energy due to contraction mechanisms. It has also been seen when muscle exercise is activated by hypoxia. Its crucial role in glucose consumption has been demonstrated physiologically and pharmacologically. This mechanism, observed *in vitro*, appeared independently of the insulin-induced mechanism. To the extent that AMPK activates glucose consumption in the muscle, blood glucose levels drop. The magnitude of this phenomenon depends on the muscle glycogen content, from the degree of training of the muscle and the type of muscle fibers it makes up. Another phenomenon involved in the consumption of glucose by AMPK is its action of it on the sensitivity of the insulin receptor, associated with the decrease in intracellular triglycerides (TG). Such TGs hinder insulin signaling and the reduction in Malonyl CoA, which are involved in the pathogenesis of IR. AMPK also stimulates the translocation of glucose transporters, Glut 4. The presence of nitric oxide stimulates the AMPK, which, in turn, stimulates glucose metabolism.

SIGNALS TOWARDS GLUCOSE TRANSPORT DURING THE EXERCISE

Exercise induces, independently of the mechanism of insulin, the uptake and consumption of glucose on the part of the muscle. One of the essential mechanisms is mediated by the variations in the concentrations of Ca2+ utilized through the activation of the PKC. It has to do with AMPK and other proteins, like p38 and the MAPK family, that are part of the insulin signaling cascade. The muscle takes in glucose during exercise, which causes the capillaries and blood flow to increase. This makes it easier for glucose to get to the cell membrane, which is made possible by the translocation pathway of the Glut 4 and the microtubules that carry it from the intracellular space to the membrane. Diacylglycerols are incorporated in the repeated contraction of the muscle, which inhibits the cascade of insulin signals and, therefore, functions as proteins that moderate the entry of glucose into the cell cytoplasm.

ASPECTS RELATED TO PHYSICAL TRAINING

Physical training increases the possibility of metabolizing or oxidizing fatty acids. This can modify the oxidative threshold, and the trained muscle can change the oxidative ratio of fats and glycogen under the same load. On the other hand, excess plasma glucose, accompanied by insulin elevations, has been observed to inhibit the oxidation of fatty acids in the muscle. This effect does not happen in reverse; increasing circulating fatty acids that do not increase the chances for oxidative degradation of fats by the muscle during exercise. This is important in facing the different states of pathophysiological factors related to IR (Fig. **4**).

GLOSSARY

Amylin: A hormone released along with insulin from beta cells that decreases glucose levels during meals.

Artificial Sweeteners: Sugar substitutes that taste sweet, have no carbohydrates and essentially no calories, and will not raise blood sugars.

Basal Infusion Profile: An insulin pump refers to the amount of insulin delivered every half hour or hour over 24 hours to provide background (*i.e.*, overnight, fasting, and between-meal) insulin replacement.

Basal Insulin Replacement: Background insulin replacement is required to stabilize the blood sugar overnight while fasting and between meals.

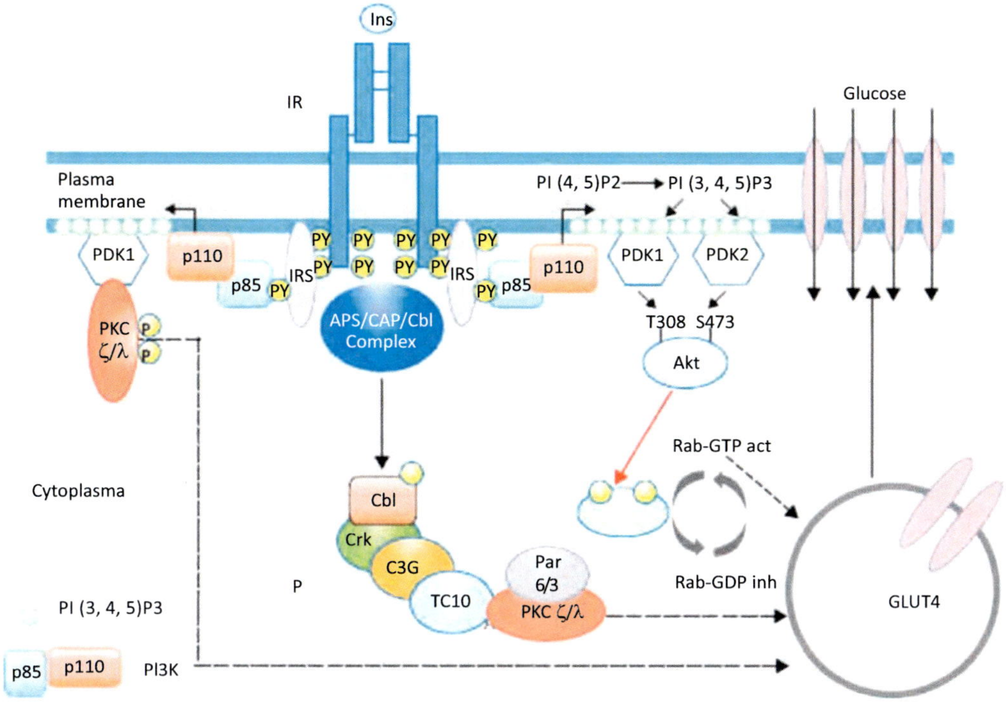

Fig. (4). Regulation of glucose transport by insulin. Insulin promotes the translocation of the Glut 4 transporter from intracellular compartments to the plasma membrane. The AS160 protein, in its non-phosphorylated and active state, negatively regulates the small G protein Rab, which participates in the vesicular Glut 4 traffic. AS160 stimulates hydrolysis of GTP bound to Rabs (generating Rab-GDP, inactive) and inhibiting vesicular traffic. When AS160 is phosphorylated by Akt and is inhibited, which increases the Rab-GTP co-dependent (active) Glut 4 traffic to the plasma membrane, on the other hand, PDK1 also induces the phosphorylation of critical sites in the activation loop of two atypical forms of PKC, which contribute significantly to insulin-induced Glut 4 translocation.

Beta Cells: Specialized cells, which make and release the hormone insulin, found in the islets of Langerhans in the pancreas.

Blood Glucose: The primary sugar is the body's fuel source. Glucose is carried through the bloodstream to provide energy to all cells in the body.

Blood Sugar Rebound: The hormonal reaction of the body to low blood sugar (hypoglycemia) that results in high blood sugar – also known as the Somogyi effect.

Continuous Glucose Sensors: Sensors that continuously measure the glucose level in the fluid between cells (interstitial fluid.) The average glucose level is displayed on a monitor.

Extended Bolus: An insulin pump term refers to a continuous delivery of bolus insulin over a certain period designated (programmed) by the user.

Fasting Glucose: Glucose level in the bloodstream after not eating for 8 to 12 hours.

Fasting Plasma Glucose Test: Test glucose level in the bloodstream after not eating for 8 to 12 hours; often used to diagnose prediabetes and diabetes.

REVIEW QUESTIONS

1. How common is diabetes?

2. Who gets diabetes? What are the risk factors?

3 What causes diabetes?

REFERENCES

[1] Krambovitis E, Porichis F, Spandidos DA. HIV entry inhibitors: A new generation of antiretroviral drugs. Acta Pharmacol Sin 2005; 26(10): 1165-73. [http://dx.doi.org/10.1111/j.1745-7254.2005.00193.x] [PMID: 16174430]

[2] Clavel F, Hance AJ. HIV drug resistance. N Engl J Med 2004; 350(10): 1023-35. [http://dx.doi.org/10.1056/NEJMra025195] [PMID: 14999114]

[3] Viganò A, Schneider L, Giacomet V, *et al.* Efficacy and tolerability of multiple drug therapy in HIV-infected children. J Infect 2005; 50(5): 404-11. [http://dx.doi.org/10.1016/j.jinf.2004.08.001] [PMID: 15907548]

[4] Geretti AM, Easterbrook P. Antiretroviral resistance in clinical practice. Int J STD AIDS 2001; 12(3): 145-53. [http://dx.doi.org/10.1258/0956462011916938] [PMID: 11231866]

[5] Kuritzkes DR. Extending antiretroviral therapy to resource-poor settings. AIDS 2004; 18: S45-8. [http://dx.doi.org/10.1097/00002030-200406003-00009] [PMID: 15322484]

CHAPTER 14

Molecular Basis of Obesity

Itrat Fatima Toor[1,*]

[1] *National Institute of Food Science and Technology, University of Agriculture, Faisalabad, Pakistan*

Abstract: Based on the classification by the World Organization of Health (WHO), it considers that a BMI equal to or greater than 30 kg/m^2 corresponds to obesity. Likewise, a BMI value equal to or greater than 25 kg/m^2 increases the chances of developing diseases associated with obesity. It is estimated that heredity in the variation of the BMI is in the range of 0.4 to 0.7; that is, the probability of inheriting obesity is very low and is more associated with exogenous factors. Obesity comorbidities are a risk factor for developing insulin resistance (IR), DM2, CVD, stroke, osteoarthritis, endometrial, breast, and colon cancer, among other chronic noncommunicable conditions. In addition, obesity is also linked to various digestive diseases, including gastroesophageal reflux disease, esophagitis, colorectal polyps, and non-alcoholic steatohepatitis. Obesity and overweight are associated with 44% of DM2 cases, 23% of ischemic heart disease cases, and 7 to 41% of cancer cases.

Keywords: Agouti-related protein, BMI, Bardet-biedl syndrome, Diabetes mellitus, Insulin resistance, Leptin, Neuropeptide Y, Prader-Willi syndrome.

INTRODUCTION

During the last decades, obesity has become a disease with a high prevalence worldwide. As of 2012, Mexico ranks first in adult obesity and second in childhood obesity. Obesity increases the risk of suffering six times from noncommunicable chronic degenerative diseases, such as diabetes mellitus type 2 (DM2), cardiovascular disease (CVD), and cerebrovascular disease. These three diseases constitute the leading causes of mortality in Mexico. To date, scientific and technological advances have provided an incomplete answer to the etiology of obesity. It is evident that genetic and environmental factors, such as an improper diet and physical inactivity, participate in developing this disease. It is essential to understand the molecular mechanisms that regulate hunger and induce food

* **Corresponding author Itrat Fatima Toor:** National Institute of Food Science and Technology, University of Agriculture, Faisalabad, Pakistan; E-mail: itratfatima71@gmail.com

Sanaullah Sajid, Sajjad ur Rahman, Shahid Mahmood, Shayan Bashir & Mudasser Habib (Eds.)

intake, as well as satiety signaling mechanisms, to find alternatives in preventing and treating obesity [1].

OBESITY

Obesity is a multifactorial chronic degenerative disease characterized by a disproportionate increase of adipose tissue, or adiposity, in the body, which is associated with poor health. This increase in adipose tissue is due to the positive energy balance resulting from an imbalance between caloric intake and energy expenditure [2]. The body mass index (BMI) is an indirect estimate to diagnose obesity and was devised by the statistician L. A. J. Quetelet. The BMI is the ratio resulting from dividing body weight in kilograms (kg) by square meters (m^2) height. The BMI is also known as the Quetelet index.

BMI = weight in kg/height in m^2

Based on the classification of the World Organization of Health (WHO), it is considered that a BMI equal to or greater than 30 kg/m^2 corresponds to obesity. Likewise, a BMI value equal to or greater than 25 kg/m^2 increases the chances of developing diseases associated with obesity. It is estimated that heredity in the variation of the BMI is in the range of 0.4 to 0.7; that is, the probability of inheriting obesity is very low and is more associated with exogenous factors. Obesity and related comorbidities are risk factors for developing insulin resistance (IR), DM2, CVD, stroke, osteoarthritis, endometrial, breast, and colon cancer, among other chronic noncommunicable conditions. In addition, obesity is also linked to various digestive diseases, including gastroesophageal reflux disease, esophagitis, colorectal polyps, and non-alcoholic steatohepatitis. Obesity and overweight are associated with 44% of DM2 cases, 23% of ischemic heart disease cases, and 7 to 41% of cancer cases.

ADIPOSE TISSUE

Adipose or fatty tissue is of mesenchymal origin and represents a connective tissue formed by adipocytes that accumulate triglycerides in their cytoplasm. These adipocytes have two primary functions: reserve of energy and thermogenesis. However, fatty tissues are currently recognized as endocrine organs because they respond to traditional hormones and the central nervous system (CNS) systems and express and secrete various proteins with endocrine functions. These include adiponectin, leptin, resistin, multiple cytokines, complement components, plasminogen activator inhibitory factor-1, and proteins of the renin-angiotensin system. Therefore, the knowledge of these molecules' function will contribute to understanding the complex development of obesity.

In the adults, the positive energy balance causes an increase in the size of the adipocyte, a well-known phenomenon as hypertrophy. In contrast, the rise in the number of adipocytes predominates in children and adolescents, known as hyperplasia. The hypertrophy of the adipocyte causes modifications in its gene expression, increasing the synthesis of tumor necrosis factor-alpha (TNFα), a proinflammatory cytokine synthesized mainly by macrophages. TNFα, in turn, increases the secretion of the proinflammatory cytokine interleukin 6 (IL6). Thus, TNFα and the protein binding to serum retinol and C-reactive protein are considered markers of inflammation in obesity. Another change in gene expression is the alteration in the synthesis of proteins such as adiponectin, leptin, and resistin.

Adiponectin

The adiponectin gene, or ADIPOQ, also known as ACDC, Acrp30, apM1, or GBP28, is located in chromosome 3q27 and is expressed mainly in white adipose tissue. Adiponectin is a hormone of 244 amino acids that contains a signal peptide, allowing its secretion outside the adipocyte. Adiponectin is post-translationally modified by hydroxylation and glycosylation. Due to these modifications, it can form high molecular weight trimers, hexamers, and isoforms. There is an inverse relationship between circulating levels of adiponectin and the amount of fat mass present in an organism or its BMI. In a longitudinal study in adults, it was suggested that low plasma adiponectin concentration is a factor of risk to present DM2. This argument was based on the fact that the subjects who developed DM2 had concentrations 22% lower than those who did not. However, in this study, a group of subjects with type II obesity had a physiological range of plasma adiponectin and did not develop metabolic comorbidities. In mouse models treated with adiponectin, this is a protective factor for IR to a greater degree in obese mice than in those with an average weight. A study on Mexican children demonstrated that adiponectin has a robust inverse association with age, BMI, and insulin concentrations. Besides, adiponectin improves insulin sensitivity, inhibits gluconeogenesis in the liver, inhibits the expression of adhesion molecules, blocks macrophage migration, prevents foam cell formation, reduces B cell production, decreases T cells' response, and favors the translocation of glucose transporters.

Leptin

The leptin gene (from the Greek leptos, meaning thin) was discovered in 1994 in the mouse. It is found on chromosome 7q31.3 in humans and has been called the LEP and OB genes. The OB gene encodes a non-glycosylated hormone-like protein of 167 amino acids, with a signal sequence of 21 amino acids removed before leptin is transported into the bloodstream as a 146 amino acid protein. In its active form, it contains a disulfide bond interchain, which is essential for the protein to exert its biological activity. Leptin presents a high homology between the different species, equivalent to 84% in the mouse and 83% in the rat. It is mainly expressed in adipose tissues, and circulating levels are proportional to fat mass. The increase in fat mass is correlated with the increase in leptin synthesis. Among the functions of leptin are decreased food intake, increased energy expenditure, lipolysis stimulation and sensitivity to insulin. For this reason, leptin levels have been related to obesity, DM2, and IR.

In animal models, it has been observed that receiving a high-fat diet increases plasma levels of leptin. Leptin crosses the blood-brain barrier through a saturable transport system so that, although leptin levels are high, not all of it will reach the CNS. This fact is known as peptidergic blockade or leptin resistance (LR). In obese people, IR and LR coexist, so it has been postulated that leptin is responsible for the relationship between obesity and IR. Children and adolescents who have DM2 and are overweight have a higher concentration of leptin concerning healthy ones. Another factor affecting leptin levels is weight loss, which leads to a rapid and disproportionate decrease in leptin. Variations in leptin concentrations also depend on sex since it has higher levels in women. The administration of exogenous leptin has been shown to normalize hyperglycemia and hyperinsulinemia in animal models of genetic obesity. The obesity accompanied by LR in humans is more severe when leptin enters the CNS; it induces the expression of cocaine and amphetamine-regulated transcript (CART) and proopiomelanocortin (POMC), and POMC, in turn, stimulates melanocortin receptors 3 (MC3R) and 4 (MC4R); this process induces satiety and blocks the hunger stimulus.

Resistin

The resistin (RETN) gene is located on chromosome 19p13 and encodes a pre-peptide of 108 amino acids, with a hydrophobic signal peptide that is cleaved before its secretion. Resistin is a hormone derived from its participation in the IR, belongs to the family FIZZ (found in the inflammatory zone), and is also known as ADSF (adipocyte-specific secretory factor). Resistin is expressed in visceral adipose tissue and macrophages, and its expression has been shown to increase during adipocyte differentiation; by decreasing visceral adipose tissue, resisting levels' decrease in plasma. It has also been shown that, in mice, resistin levels are related to blood glucose. The role of protein in humans is controlled since not all people with obesity and DM2 have elevated plasma levels of resistin. Some investigators support the idea that resistin single nucleotide polymorphisms are associated with the development of obesity, IR, and DM2. It has been reported in humans that the G genotype of the -420G promoter region increases gene expression and plasma resistin levels, related to the development of obesity and IR. However, another group of researchers found no correlation between plasma resistin levels and SNPs in obesity or DM2. Because macrophages mostly make resistin and its levels are linked to IL-6 levels, it has been shown that it is closely linked to an inflammatory environment (Fig. **1**).

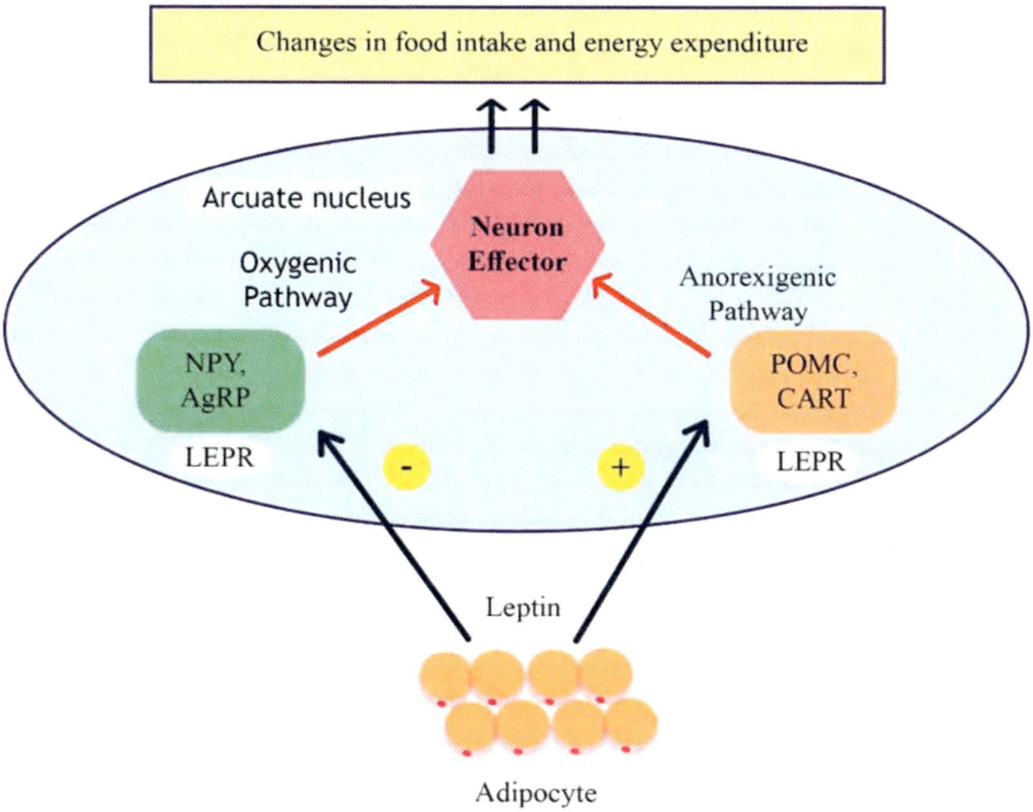

Fig. (1). Leptin binds to the leptin receptor which is found on the surface of secretory cells proopiomelanocortin (POMC) and cocaine and amphetamineregulated transcript (CART); in turn communicates with a neuron effector, decreasing food intake. The Leptin release from adipose tissue is dependent on adiposity.

REGULATION OF HUNGER AND SATIETY CENTRAL REGULATION

For half a century, various physiological mechanisms have been involved in the regulation of homeostasis of body weight; one of these mechanisms is carried out by leptin, which acts in the CNS. Adiposity causes increased leptin synthesis; leptin enters the CNS towards the arcuate nucleus and stimulates the neurons producing POMC and CART. At the same time, leptin acts on neurons that synthesize neuropeptide Y (NPY) and Agouti-related protein (AgRP), blocking their synthesis [3].

POMC is a 285 amino acid protein synthesized in the arcuate nucleus and solitary tract. This protein can undergo proteolytic cleavage resulting in the synthesis of eight different peptides, as the three isoforms alpha, beta, and gamma melanocyte-stimulating hormone. CART is a protein that promotes negative energy balance, suppresses food intake and exerts its effects by binding to MC3R and MC4R, which are expressed mainly in the brain. POMC and CART will communicate with an effector or second-order neuron and inhibit food intake through this series of signals called the anorexigenic pathway. On the other hand, NPY and AgRP are stimulators of food intake and conform to the orexigenic pathway. NPY is an anabolic signaling peptide that stimulates hunger or appetite. It is synthesized in the body cells of neurons from the arcuate nucleus in the hypothalamus and transported axonally to the paraventricular heart, located in high concentrations. On the other hand, AgRP is an antagonist of MC3R, so it also stimulates hunger.

Peripheral Regulation

Peptides and Hormones of the Gastrointestinal Tract

The regulation of short-term intake involves neural and endocrine signaling. The gastrointestinal tract has mechanoreceptors and chemoreceptors that carry the stimulus from the vagus nerve to the medulla oblongata and generate satiety signals. On the other hand, gastrointestinal hormones stimulate ascending vagal pathways that go from the gastrointestinal tract to the spinal bulb or directly to neurons in the hypothalamus. The presence of peptides and hormones that act as hunger and fullness signals from the gastrointestinal tract makes you feel full and reduces the amount of food you eat in a certain amount of time.

Preproglucagon and its Derivative

The preproglucagon gene is located on chromosome 2q36-37, and its product is a preproprotein modified to give rise to four mature peptides: peptide YY, glucagon-like peptide-1 (GLP-1), oxyntomodulin and glucagon-like peptide-2.

Peptide YY

It is a peptide of 36 amino acids whose main isoform is truncated and is known as PYY3-36. It is synthesized in the small intestine's L cells and found in the highest concentration in the colon and rectum. PYY3-36 is released into the circulation after ingesting food, and its level is reduced by fasting, and its administration causes a reduction in food intake. This effect can result from activating the route to reduce the food intake or because it triggers a consequence of the refusal of food represented as nausea. However, in some reports, very low circulating levels of PYY have been observed in obese people in fasting.

Glucagon-like Peptide

GLP is a neuropeptide hormone synthesized by posttranslational processing of preproglucagon in the CNS and the intestine. This hormone activates neurons in the arcuate nucleus, the paraventricular hypothalamic nucleus, the nucleus of the solitary tract, and the area postrema. It is found in greater amounts in the medulla oblongata, which contributes to energy homeostasis. GLP-1 is released into the circulation after food intake, and it induces satiety and increases spending energy by increasing body temperature. In addition, it works as a suppressor of the secretion of stomach acids.

Oxyntomodulin

Oxyntomodulin reduces food intake when administered to humans by peripheral pathways and promotes energy expenditure. It comes from the preproglucagon gene and enters the bloodstream after eating food. Oxyntomodulin sticks to the GLP-1 receptor (GLP1-R). The medulla oblongata and the arcuate nucleus are where these receptors were found.

Cholecystokinin

The cholecystokinin (CKK) gene, called CCK, is located on chromosome 3p22-21.3 and is widely expressed in the gastrointestinal tract, particularly in the duodenum and jejunum. It is a protein consisting of 115 amino acids, and it is released when lipids come into contact with the stomach. A part of the CCK acts located on paracrine stimulating CCK1 receptors on sensory fibers of the vagus nerve. Another aspect of the secreted CCK is that it enters the blood circulation and stimulates the pancreas and gallbladder functions. It is also expressed in the CNS and acts as a neurotransmitter regulating reward behaviour, satiety, memory, and anxiety. CCK rapidly reduces the amount of food eaten and the mealtime duration, and bloating potentiated gastric effects. In addition, it slows gastric emptying and increases intestinal motility. The long-term impact of CCK on body weight may be a partial result of interaction with molecules such as leptin, which enhance the satiety effect of CCK.

Bombesin

Bombesin is a tetra decapeptide, the structure of which is similar to that of gastrin-releasing peptide (GRP) and that of neuromedin B. Bombesin is widely synthesized in the intestine, and its plasmatic levels increase after eating, which also, reduce food intake regardless of the CCK. The glucose-dependent insulinotropic polypeptide (GIP) is a 42 amino acid peptide released by K cells of the duodenum after food intake. It has not yet been reported to influence food intake, but resistance to obesity has been observed in the GIP receptor knockout in mice fed a high-fat diet. Therefore, it is suggested that it directly affects adipocytes rather than affecting the central regulation of hunger and satiety.

Ghrelin (Orexigenic Hormone)

The ghrelin and obestatin prepropeptide (GHRL) is located on chromosome 3p26-p25. The product of this gene is processed post-translationally to produce ghrelin and obestatin. Ghrelin is a peptide hormone of 28 amino acids, synthesized mainly by cells oxyntic in the stomach and the duodenum, the ileum, cecum, and colon. It is considered the only peripherally active orexigenic hormone, and its participation in the stimulation of hunger seems to be mediated by the hypothalamus. In the long term, ghrelin levels are regulated by energy homeostasis. When energy intake is balanced with energy expenditure, it restricts ghrelin secretion. Still, if the energy reserves decrease rapidly to a specific limit, this change causes an increase in its release. In the short term, the circulating ghrelin concentration rises with fasting and is reduced after eating food, so ghrelin is considered to be involved in obesity. The Ghrelin administration in humans increases the sensation of hunger and food intake; it also promotes fat storage in adipose tissue. The other product of the GHRL gene is the peptide called obestatin, which binds to the orphan receptor 39 coupled to G proteins (GPR-39) in the stomach, jejunum, and ileum. The administration of obestatin *via* the peripheral or intraventricular route in murine models reduces food intake and, in addition, *via* the peripheral route, it also causes a decrease in weight gain.

Pancreatic Hormones

Pancreatic polypeptide (PP) consists of 36 amino acids, which the pancreas releases after food intake. It is derived from a precursor called pancreatic pre-pro polypeptide, whose gene is located on chromosome 17q21. The amount of PP released is proportional to the energy content of the food consumed. In addition, PP is released and subsequently transported to circulation, and lipid digestion is required. Intravenous infusion of PP reduces food intake in the short term (2 hours) and the following 24 hours. This effect may be due to delayed gastric emptying [4].

Amylin

Amylin, also known as an amyloid polypeptide of the islets, is a residual peptide of 37 amino acids that belong to the family of calcitonin peptides, each encoded by the islet amyloid polypeptide gene (IAPP), which is located on chromosome 12p12.1. Amylin is released along with Insulin by pancreatic β cells in response to food intake. Although it is believed that its primary function is to participate in glucose homeostasis, administering peripherally causes weight reduction; if administered at supraphysiological doses, it may reduce food intake.

Insulin

Insulin is considered a critical peripheral regulator of food intake. In humans, the insulin gene (INS) is located at 11p15.1 and codes for prepro Insulin, giving rise to Insulin. The plasma levels of Insulin directly correlate with body weight and adiposity. Plasma insulin is increased with food intake and in any other condition of positive energy balance, it decreases during fasting and in periods of negative energy balance. Insulin correlates with long-term energy balance; on the other hand, the insulin receptor is expressed at the central level in the arcuate, dorsomedial, and paraventricular nuclei, so plasma insulin levels participate in energy homeostasis. Evidence shows that insulin acts as an anorexigenic signal in the central nervous system. The central administration of insulin decreases food intake and body weight but also alters gene expression of the hypothalamus known to regulate food intake. Recently, the melanocortin system has been discovered to be a mediator of the central actions of Insulin.

FACTORS ASSOCIATED WITH OBESITY

The factors that affect the corticolimbic areas of the brain are related to learning, memory, reward, humour, and emotions; as a consequence, positive associations are created between food and psychological factors that contribute to the cancellation of organic signals, both hunger and satiety. Even other factors, such as customs, habits, adherence to schedules and stress, generate changes in food. The facts mentioned cause excessive consumption of nutrients when they are not needed. On the other hand, physical inactivity is a consequence of the change in working hours, the lack of spaces conducive to exercise, the decrease in the need to travel to meet basic needs, *etc*. Physical inactivity or a sedentary lifestyle leads to the reduction in energy expenditure and to the fact that energy reserves are not used. Thus, environmental factors contribute to the imbalance between caloric intake and energy expenditure, promoting obesity [5].

GENETICS OF OBESITY

Monogenic Obesity

Monogenic obesity is due to the mutation of a single gene regulating hunger and satiety. It should be noted that these types of modifications are infrequent and cause severe obesity. Mutation homozygosity of the LEP gene causes congenital leptin deficiency, which causes hyperphagia and leads to early-onset Obesity and severe Obesity in adulthood. Heterozygous individuals for the said mutation have lower circulating leptin levels and a higher prevalence of obesity. On the other hand, homozygous mutations of the leptin receptor gene generate a truncated receptor before the transmembrane domain, causing a phenotype similar to that of congenital leptin deficiency. Homozygous POMC mutations cause total loss of function, or heterozygous mutations lead to hyperphagia and obesity at the early start. Another gene with heterozygous mutations described is prohormone convertase 1 (PC-1), responsible for posttranslational modifications of proteins; for example, POMC, one of the leading monogenic causes of obesity is MC4R deficiency, whose heterozygous mutations have been reported in various ethnic groups.

Syndromic Obesity

Genetic defects or chromosomal abnormalities cause at least 20 autosomal and X-linked syndromes. These syndromes are characterized by presenting obesity and, in the majority, mental retardation. Prader-Willi syndrome (PWS) is one of the most frequent autosomal diseases characterized by obesity, hyperphagia, hypotonia, mental retardation, short stature, and hypogonadism. Recently, it was suggested that elevated ghrelin levels in people with PWS could increase the appetite for interaction with POMC/CART and NPY. In humans, deletion or disruption of SIM1 (single-minded homolog 1) results in a similar phenotype to that of PWS or in early development of obesity associated with hyperphagia. Bardet-Biedl syndrome (BBS) is characterized by cone-rod dystrophy in the eye, polydactyly, learning difficulties, hypogonadism in males, renal malformations, and obesity that develops in the first five years of life.

Polygenic Obesity

There is increasing evidence that genetic variations, such as SNPs, play an essential role in the obesity epidemic. These SNPs have low effects on individual susceptibility to standard forms of obesity but, due to their high frequency, they may contribute more to obesity on a large scale. For the study of genetic variations, strategies have been developed that allow us to identify the genes involved in polygenic obesity, such as the study of candidate genes or whole genome sequences to detect chromosomal regions linked to quantitative traits associated with obesity. Most genes regulate energy metabolism, appetite control, or autocrine and paracrine signaling. Obesity related to various SNPs is the most frequent disease. SNPs have been detected in lipid and glucose metabolism genes and genes that regulate hunger, satiety, and energy expenditure.

GLOSSARY

Obesity: A medical condition characterized by excessive accumulation of body fat leading to various health complications.

Adipose Tissue: A type of connective tissue that stores energy in the form of fat and serves as an endocrine organ.

Adipocyte: A cell that stores fat in adipose tissue.

Adipokines: Hormones and cytokines secreted by adipose tissue that regulate energy metabolism and inflammation.

Energy Metabolism: The process by which the body converts food into energy for cellular processes.

REVIEW QUESTIONS

1. What does adipose tissue express the proteins?
2. What factor determines the modification of the levels of leptin?
3. Mention the effects of leptin at the central level on the orexigenic and anorexigenic pathways.

REFERENCES

[1] Aguilar-Salinas CA, García EG, Robles L, *et al.* High adiponectin concentrations are associated with the metabolically healthy obese phenotype. J Clin Endocrinol Metab 2008; 93(10): 4075-9. [http://dx.doi.org/10.1210/jc.2007-2724] [PMID: 18682512]

[2] Bell CG, Walley AJ, Froguel P. The genetics of human obesity. Nat Rev Genet 2005; 6(3): 221-34. [http://dx.doi.org/10.1038/nrg1556] [PMID: 15703762]

[3] a. Benoit SC, Clegg DJ, Seeley RJ, Woods SC. Insulin and leptin as adiposity signals. Recent progress in hormone research. 2004 Jan 1;59:267-86. b. Canizales S. Genetic aspects of obesity. Rev Endocrinol Nutr 2008; 16(1): 9-15.

[4] de Graaf C, Blom WAM, Smeets PAM, Stafleu A, Hendriks HFJ. Biomarkers of satiation and satiety. Am J Clin Nutr 2004; 79(6): 946-61. [http://dx.doi.org/10.1093/ajcn/79.6.946] [PMID: 15159223]

[5] Gómez R, Conde J, Gómez Reino JJ, Lago F, Gualillo O. Adipocytokines: Emerging mediators of the immune response and inflammation. Reumatol Clin 2009; 5: 6-12. [PMID: 21794635]

CHAPTER 15

Molecular Basis of Hepatitis B

Saima Pervaiz[1,*] and **Sara Masood Cheema**[2]

[1] *Department of Immunology, University of Health Sciences, Lahore, Pakistan*

[2] *Department of Pathology, Azra Naheed Medical College, Superior University, Lahore, Pakistan*

Abstract: Hepatitis B virus (HBV) is a serious global health problem affecting millions worldwide. Chronic HBV infection can lead to liver cirrhosis and hepatocellular carcinoma, making it a major cause of morbidity and mortality. The molecular basis of HBV infection and pathogenesis is complex and involves multiple interactions between the virus and the host immune system. HBV is a partially double-stranded DNA virus replicating through reverse RNA intermediate transcription. The virus has several proteins, including the envelope protein (HBsAg), core protein (HBcAg), and polymerase (HBp), that play critical roles in virus entry, replication, and assembly. The viral genome is organized into four overlapping open reading frames (ORFs), each encoding a different viral protein. During HBV infection, the virus initially binds to heparan sulfate proteoglycans on the cell surface, followed by binding to specific receptors, such as the sodium taurocholate co-transporting polypeptide (NTCP). The virus then enters the cell through endocytosis, where it is uncoated and releases the viral DNA into the nucleus. The viral DNA is then transcribed by the host RNA polymerase II, producing viral mRNAs translated into viral proteins. One key factor determining the outcome of HBV infection is the host's immune response. The innate immune response plays an important role in controlling the initial phase of HBV infection, while the adaptive immune response, particularly the CD8+ T cell response, is critical for the clearance of the virus. However, in some cases, the immune response cannot clear the virus, leading to chronic infection. Understanding the molecular basis of HBV infection and pathogenesis is critical for developing effective treatments and vaccines. Current treatments for chronic HBV infection include nucleoside/nucleotide analogs and interferon-based therapies, which can suppress viral replication and reduce liver damage. However, these treatments are not curative and can have significant side effects. Vaccination against HBV is highly effective in preventing infection and is recommended for all individuals at risk of HBV infection.

Keywords: Adaptive immune response, Core protein, CD8+ T cells, Cell surface receptors, Envelope protein, Hepatitis B virus, Innate immune response, Nucleoside/nucleotide analogs, Liver cirrhosis, RNA intermediate.

* **Corresponding author Saima Pervaiz:** Department of Immunology, University of Health Sciences, Lahore, Pakistan; E-mail: saimapz@hotmail.com

Sanaullah Sajid, Sajjad ur Rahman, Shahid Mahmood, Shayan Bashir & Mudasser Habib (Eds.)

INTRODUCTION

The term hepatitis means "inflammation of the liver", and the most common causes of this inflammation are infections by one or more of the viruses known as hepatitis A, B, C, D, and E viruses (HAV, HBV, HCV, HDV, HEV, respectively). These viruses can cause an acute illness with symptoms lasting several weeks, such as jaundice (yellowing of the skin and eyes), dark urine, extreme fatigue, nausea, vomiting, and abdominal pain. Of the viruses mentioned above, HBV and HCV can be asymptomatic, with no apparent liver damage, in carriers. They can also cause clinical manifestations, such as the onset of chronic hepatitis with progression to liver cirrhosis, which may culminate in hepatocellular carcinoma.

In relation to HBV, advances in molecular biology have contributed considerably to preventing, diagnosing, and treating hepatitis B. The genomic organization and mechanisms of transcription and replication of the virus are now well understood. For this purpose, diagnostic methods are available that make it possible to directly identify the presence of the genome of this virus in biological samples. In addition, a highly effective recombinant vaccine is also available (offers 90% protection) with minimal side effects. The production of recombinant antigens has also improved serological diagnosis for detecting viral antigens and/or antibodies formed in response to infection. Despite the efforts made by different research groups, the ideal treatment has yet to be available. The drugs approved to treat hepatitis B are interferon, lamivudine, adefovir, dipivoxil, entecavir, and tenofovir. Prospective targets and ongoing investigations include agents that block the NTCP receptor and the development of short-interfering RNAs that may disrupt the transcription and translation of HBV DNA. There are also possibilities of developing immunomodulatory vaccines by including additional Pre S antigens into routine Hepatitis B vaccines. Antiviral therapy has shown a clinical improvement in some patients and a decrease in the occurrence of hepatocellular carcinoma in countries where this condition is highly endemic. However, drug-resistant variants of the virus are commonly present, highlighting the need to extend prevention and diagnostic measures worldwide and continue the search for new and more effective treatment strategies to eradicate hepatitis B [1].

GLOBAL EPIDEMIOLOGY

The World Health Organization (WHO) estimates that approximately 2 billion people are infected with HBV, of which 350 million are chronically infected. HBV is the tenth leading cause of death worldwide, and about 500,000 to 1.2 million people die each year from complications of HBV infection. The distribution of HBV infection in the world is heterogeneous, so it is divided into three geographic categories based on the prevalence of chronic HBsAg carriers:

(a) areas with a high prevalence (>8%); (b) areas with intermediate prevalence (2 to 8%), and c) areas with a low prevalence (< 2%). Among the regions with high prevalence are some African, Southeast Asia, and China countries. Areas of intermediate prevalence include Eastern and south-eastern Europe, North Africa, the Middle East, Japan, India, and parts of South America. The United States and Western Europe are low-endemic areas. In Mexico, the studies, although not representative of the general population, suggest that it is a low-endemic country with some regions of intermediate prevalence.

Routes of Transmission

- Perinatal (Transmission takes place during birth. Transplacental transmission is very unlikely and so far there is no evidence of transmission due to breastfeeding).
- Unsafe injections and transfusions of blood products.
- Use of non-sterile syringes and needles.

In high-prevalence countries, most infections are acquired perinatally and during early childhood, when the immune system is immature. In these regions, hepatocellular cancer is among the three leading causes of death. In countries with moderate and low endemicity, hepatitis B is acquired in young adults through sexual activity and intravenous drug use. In addition, hepatitis B is the leading viral infection that occurs in healthcare workers through an accidental puncture or by working without sufficient preventive measures, especially in countries where vaccination is not widely used.

HEPATITIS B VIRUS

To understand the concepts of the natural history of the hepatitis B virus, its prevention, diagnosis, and treatment, first of all, it is essential to know the fundamental aspects of HBV, its genome organization, and the most important characteristics of its proteins. HBV is classified within the family Hepadnaviridae of the Orthohepadnavirus genus, being tropic for the liver, and is characterized as a double-stranded circular DNA virus replicating through RNA intermediary by reverse transcription. The HBV genome can integrate into host chromosomes [2].

In vivo, the liver is the main organ of virus replication; however, viral genomes have been found outside the liver, as in the bile duct epithelium, the pancreas, renal cells, lymphocytes, and monocytes. The mature viral particle comprises a nucleocapsid surrounded by a lipid bilayer into which the envelope proteins are incorporated. Inside the nucleocapsid is the viral genome containing all the organized genetic information.

GENOMICS AND PROTEOMICS

The HBV genome consists of DNA of 3,200 bp, although its size varies according to the genotype. This peculiar characteristic of having a small genome is added that during replication, it uses an intermediate RNA (with a reverse transcriptase that does not correct errors); it can be concluded that it is a virus with an intermediate mutation rate (between a DNA virus and an RNA virus). A further characteristic of HBV DNA is that one of its strands is smaller than the other. It is divided into four main regions: open-reading fragments (ORFs): S, C, P, and X. These ORFs contain the genes that code for the viral envelope proteins, capsid proteins, the E antigen (HBeAg), polymerase, and the X protein. Since these genes overlap, most of the DNA bases of the HBV genome are involved in coding two viral proteins. Table **1** shows the HBV genome's organization, transcripts, proteins, and the main characteristics of these molecules.

Table 1. HBV genome and proteome and main characteristics of its molecules.

ORF	Gene/Region	Encoded Protein	Function and Characteristics
P	P	Polymerase	Replication of the viral genome. It contains four domains: 1) N-terminal, which participates in the encapsulation and initiation of negative-strand synthesis; 2) RT, catalytic region that synthesizes the viral genome; 3) RNase H, its function is to degrade the viral pgRNA; and 4) a spacer of unknown function.
S	Pre-S1 Pre-S2 Gen S	Three viral envelopes: pre-S1 + Pre-S2 + S gene = "Large protein. large" Pre-S2 + Gen S = "Medium Protein medium" Gene S = "Small protein."	They form a structural part of the virus envelope. The three proteins have a common amino acid sequence of the carboxyl-terminal end (corresponding to the small protein known as the HBsAg). The large protein contains a hepatocyte recognition site, which is considered to serve as an entry of HBV into the host cell. HBsAg contains the main antigenic determinants, which are recognized by the antibodies produced in response to infection or vaccination that inhibit the virus's entry into the hepatocyte. The vaccine against this virus is the first recombinant. To date, a vaccine to be produced continues to be used HBsAg as an immunogen.
C	Pre-C Gen C	Two proteins: Pre-C region + Gene C = HBeAg Gene C = The core protein (HBcAg)	HBeAg. It is the soluble form of HBcAg. Necessary for persistent infection. Crosses placental barrier. Core protein. Forms a structural part of the nucleocapsid of the virus. Approximately 180 subunits of this protein are assembled to form this viral structure. It favors the recognition of polymerase and the intermediate viral RNA, which are encapsulated and synthesized from the intermediate RNA.
X	Gen X	Protein X	Transcriptional regulator promotes the transcription of its gene and host genes, including c-myc and c-fos.

REPLICATION

Once the virus penetrates the hepatocyte, it releases the nucleocapsid into the cytoplasm. The viral genome is then transported to the nucleus by mechanisms that still need to be well-defined. Host RNA polymerase II transcribes the negative strand of the viral DNA into a 3.5 kb pregenomic RNA (pgRNA) and three messenger RNAs (mRNAs) that serve to translate viral proteins. The pgRNA leaves the nucleus and, in the cytoplasm, has two functions: on the one hand, it serves as a messenger for synthesizing the polymerase and core protein, and on the other, it is an intermediate in the replication process. If the viral DNA continues long enough in the cell, it becomes double-stranded and, in this case, returns to the nucleus for another round of replication. However, when a partially double-stranded DNA is formed, the nucleocapsid matures and interacts with envelope proteins in the endoplasmic reticulum to form the complete 42-nm viral particle, secreted by exocytosis (Fig. **1**).

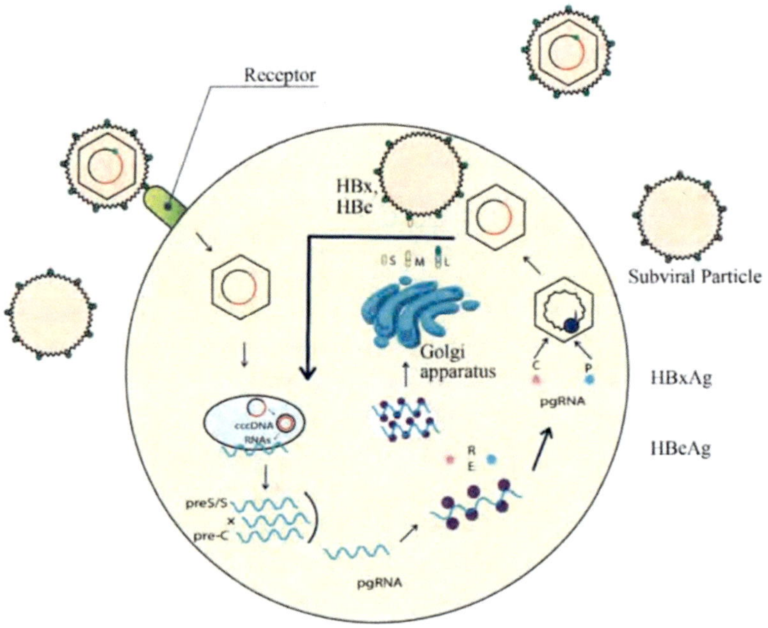

Fig. (1). HBV life cycle. HBV enters the cell through a receptor and releases the envelope into the cytoplasm; subsequently, it loses the nucleocapsid and enters the nucleus. The incomplete chain is synthesized to form the cccDNA. The RNA pol II of the host synthesizes the viral RNAs. In the free ribosomes, viral polymerase and capsid proteins are synthesized, and within the capsid, DNA is synthesized from the pgRNA. In the rough endoplasmic reticulum, proteins that envelop the viral genome are synthesized and, together with host lipids, envelop the capsid to release the complete viral particle, HBeAg and HBxAg, secreted from the cell.

NATURAL HISTORY OF HEPATITIS B

The likelihood of developing acute infection with the characteristic symptomatology (jaundice, abdominal pain, dark urine, fever) and subsequent clearance of HBV is age-related; in children under four years, it occurs in only about 10% of cases. In young adults, the proportion is higher (approximately 90%). In contrast, the likelihood of HBV chronic hepatitis is high in individuals infected perinatally (90%) or during childhood (20 to 30%), compared with the probability of being acquired when an immunocompetent adult is infected (< 10%). Three phases of chronic hepatitis have been described as observed during the follow-up of individuals infected at an early age. These phases are: a) immunotolerance, with standard liver biopsy or with minimal inflammatory changes without fibrosis; b) immunoreactive phase, which is subdivided into b1) HBeAg positive and b2) HBeAg negative, with hepatic inflammation with or without the presence of fibrosis determined by liver biopsy, and c) inactive hepatitis B phase, in which liver inflammation with or without fibrosis may improve over time and, on some occasions, HBsAg may not be detected.

Most HBV carriers experience HBeAg seroconversion; the HBe HBV protein is not detected in the blood, and antibodies directed against this protein appear. This seroconversion is generally associated with a good prognosis in the evolution of hepatitis B. The long-term outcome of chronic hepatitis B is heterogeneous. The annual incidence of clinically diagnosed cirrhosis is estimated to be 2 to 3%; risk factors for developing cirrhosis include advanced age, presence of HBeAg, and elevated alanine aminotransferase (ALT) levels. The annual incidence of hepatocellular carcinoma (HCC) is estimated to be less than 1% in patients without cirrhosis and 2 to 3% in patients with this disease.

The HCC occurs more often in men compared to women 3:1 to 4:1, and other risk factors for the development of HCC include the presence of cirrhosis, advanced age, family history of HCC, and co-infection with hepatitis C. Annual spontaneous HBsAg clearance occurs in 0.5 to 0.8% of patients with chronic hepatitis B. People who clear this antigen are considered to be in the "recovery" phase of hepatitis; however, there is evidence that DNA can be detected in blood samples even five years later, and even more so in individuals who are considered to have eliminated HBV-DNA cccDNA is still detected in the liver. Patients who have cleared HBsAg without developing cirrhosis are also at risk of developing HCC. This may occur because the HBV genome remains in very low concentrations in the blood and liver and can integrate into the hepatocyte genome, leading to cellular alterations. For this reason, people who were HBsAg and in the "recovery" phase should continue to undergo lifelong clinical assessments to detect HCC (Table **2**).

Table 2. Interpretation of serological tests.

Phase of the Disease	HBsAg	Anti-HBs	HBeAg	Anti-HBe	Anti-HBc	HBV-DNA Copies/ml	ALT	History
Acute hepatitis B	+	-	+	-	IgM+	Positive	High	Evidence of recent exposure
Resolved infection	-	+/-	-	+/-	+/-	ND	Normal	Past exposure.
Immunized	-	+	-	-	-	Negative	Normal	Vaccination.
Phases of Chronic Hepatitis B								
Immunotolerant	+	-	+	-	+	>107	Normal	Young age (2nd or 3rd decade), infected in childhood.
Immunoreactive (HBeAg +)	+	-	+	-	+	>105	High	Distant infection.
Immunoreactive (HBeAg -)	+	-	-	+	+	>104	Persistently high or indeterminate	Distant infection (more common in older people).
Inactive carrier HBsAg	+	-	-	+	+	ND-104	Normal	

Factors Influencing the Severity of Hepatitis B [3]

Viral Load

There is evidence that cirrhosis increases in patients with a high viral load.

Hepatitis B Virus Genotype

To date, ten genotypes have been defined and designated by the alphabet's letters, from A to J, and multiple subtypes. The classification is based on an 8% difference in the nucleotide sequence of the complete genome. The different genotypes have a different geographic distribution: genotype A predominates in Northern Europe, America, and India; genotypes B and C in Asia and the Pacific; genotype D is mainly detected in the Mediterranean., most of Russia, and India; genotype E is localized in Africa; genotype F in Central and South America, and genotype H is the predominant genotype in Mexico. Genotype G has been identified in France, Germany, the United States, and Mexico. Genotyping was found to help determine the prognosis of treatment response. Studies comparing genotype A with genotype D and genotype B with genotype C show that patients infected with genotypes A and B are more responsive to interferon-alpha therapy. Studies suggest that patients infected with genotype A more easily clear HBsAg without needing treatment.

Thus, in a patient with genotype A who is observed to have decreasing levels of viral DNA, it is recommended to wait to assess whether viral clearance occurs without needing treatment. On the other hand, if the patient fails to clear the viral infection through the immune system, it will be assessed whether they can receive treatment; otherwise, severe liver disease would be predicted. Genotyping could be used to identify patients at high risk of disease progression. There are studies available that indicate that there is a correlation between genotype and severity of liver disease. Infections with genotypes C, F1, and B2-5 have been associated with a high risk of developing hepatocellular carcinoma; in contrast, infections with genotypes B1, B6, and A2 appear to be associated with a lower risk of liver complications. However, the number of studies needs to be increased, as this comparison has yet to be studied in most HBV genotypes.

In Mexico, where the H genotype predominates, there are no clinical studies to determine whether there is an association between the HBV genotype and the level of severity of liver damage, response to treatment, and the vulnerability of the virus to develop resistance mutations to nucleoside and/or nucleotide analogs used in hepatitis B treatment.

Mutations of the Hepatitis B Virus and their Clinical Significance

As noted above, HBV replicates *via* an intermediate RNA (which requires a reverse transcriptase that cannot proofread errors), thus presenting a mutation rate ten times higher than other DNA viruses. This high mutation rate leads to viral variants with clinical implications. An important example is mutations in the YMDD motif of the HBV polymerase which results in resistance against Lamivudine. This is usually due to monotherapy.

Mutations Present in the Precentral and Central Region

As mentioned above, this region encodes the core protein for the HBeAg. These two proteins contain potent antigenic determinants that evoke an immune response during the mechanism of viral removal. Two mutations affect the synthesis of HBeAg: 1) those found in the precentral region and 2) in the promoter central basal. The presence of these mutations explains that the pressure exerted by the immune system to eradicate the virus leads to the development of these "escape mutations," which allows the survival of the variants of the virus that do not produce HBeAg or produce it at very low levels, but even with these mutations, the virus does not lose the ability to replicate itself. Sometimes it does it more effectively (Mutations in the central basal promoter). These mutations occur in all genotypes, but they are more common in the genotypes associated with the development of cirrhosis decompensated or with HCC, such as genotype C.

Mutations in the Pre-S/S Region

In areas endemic to hepatitis B, escape mutations to the vaccine alter the HBsAg. In general, substituting an amino acid glycine with asparagine at position 145 of the HBsAg is sufficient for the antibodies generated in response to a wild virus or the vaccine, which do not detect this antigen. Therefore, these mutations can result in a negative serological test for detecting HBsAg (hidden hepatitis B) and a poor response to the vaccine. The 145th amino acid of HBsAg is located within the determinant "a", the name given to the amino acids located from position 99 to 170 of the HBsAg. This determinant represents the main immune target of the antibodies directed against this antigen, either by an active infection or in response to the vaccine. In patients with hidden hepatitis B, mutations in the Pre-S1 and Pre-S2 regions have also been found, with nonsense deletions or mutations in the pre-S2 start codon (which affects the synthesis of the median protein envelope).

Evaluation of the Patient with Hepatitis B

The evaluation of a patient diagnosed with hepatitis B should be directed to the monitoring of the activity of the disease and the knowledge of its evolution according to what is known of the natural history of hepatitis B to evaluate the treatment and/or management of the patient appropriately. In addition to the medical history and clinical evaluation, it requires serological, biochemical, and molecular diagnosis as a virological test to confirm the diagnosis; also, imaging studies are needed to continue patient monitoring. Although it is unnecessary in all cases, liver biopsy can be very useful in providing a prognosis of the disease and offering a follow-up suitable for the selected patients for treatment. The frequency of clinical monitoring in the patient should be determined according to the activity and stage of the disease. The search aimed at the development of hepatocellular carcinoma should be based on age, sex, race, family history, duration, activity, and stage of the disease.

SEROLOGICAL DIAGNOSIS

The detection of HBsAg, the primary serological marker of screening for hepatitis B carries out the diagnosis of hepatitis B. The presence of this marker can be detected both at the acute stage of infection and in the chronic phase. The specific serological marker of acute hepatitis B is the M-type antibody directed against the central HBV protein (IgM-anti- HBC).

HBeAg in serum indicates an increased viral replication in the liver. The "seroconversion of the HBeAg" is associated with a clinical improvement of hepatitis (reduction of viral DNA, normal levels of ALT, and mild liver inflammation). This clinical improvement is accompanied by the seroconversion of the HBsAg (HBsAg negative/anti-HBs positive). The presence of the anti-HBs, in the absence of the rest of the serological markers of HBV, indicates that the person has been vaccinated against the virus or was infected but has a high probability of eliminating the virus.

Molecular Diagnostics

Four types of molecular assays are available for the diagnosis and management of hepatitis B:

1. Detection of the viral load by quantitative tests.
2. Determination of the genotypes.
3. Antiviral resistance tests to detect mutations.
4. Tests for detecting mutations in the central promoter region and the precentral.

Tests that determine the viral load of the HBV in peripheral blood (serum or plasma) are those used more often. The other tests are very specialized, so their use is more limited.

Detection of Viral Load

It is fundamental in the initial evaluation of chronic hepatitis B and during its handling, in particular, to initiate treatment and monitoring. Susceptible viral load levels are essential in diagnosing patients with low viral load, such as those who have chronic hepatitis and are HBeAg negative, like those with hidden hepatitis B. One of the fundamental aspects is the analytical sensitivity of the molecular test (minimum amount of DNA that is capable of detecting) and the dynamic range (the minimum and maximum that is detected).

Determination of Genotypes

Molecular tests for determining the genotype are helpful since the genotype could influence the symptoms of chronic hepatitis B and the success of the antiviral treatment.

Antiviral Resistance Tests

Commercial tests are available that detect the resistance to antivirals by PCR and nucleic acid hybridization. However, in most cases, these tests are not routine and only performed in research protocols. For the follow-up of the treatment of hepatitis B, viral load is routinely measured, as a 10-fold increase in such a load may indicate the presence of a mutation that confers antiviral resistance in use. The detection of antiviral resistance mutations is very valuable in the decision of the drug to be used as a treatment since there are HBV mutations that have cross-reactivity with different antivirals; that is, the same mutation affects the efficiency of other drugs. Thus, for example, the rtL180M mutation present in the reverse transcriptase domain of polymerase viral, which at position 180 changes a Leu to a Met, creates resistance of the virus to treatment with lamivudine and Telbivudine, and decreases the effect of Entecavir [4].

Tests for the Detection of Mutations in the Central Promoter and the Precentral Region

These tests are carried out on patients with hepatitis. They are negative to the HBeAg and are applied only in research projects. Soon, these tests may be very useful in the treatment, prognosis, and clinical evolution of hepatitis B.

FINAL CONSIDERATIONS

Despite the advances made in the knowledge of hepatitis B, which have led to a decrease in cases of HCC, as well as to an increase in the average and the quality of life of infected patients, it still requires a more significant number of studies, in different races, aimed at a deeper understanding of the association of the virus and host genetics, focused on knowledge from the pathophysiology of hepatitis B and to the development of new drugs more suitable to eradicate this infection.

GLOSSARY

B Surface Antigen (HBsAg): HBV envelope protein often produced in excess and detectable in the blood in acute and chronic HBV infection.

HB Core Antigen (HBcAg): HBV core protein. The core protein is coated with HBsAg and, therefore, is not serum-free.

HB e Antigen (HBeAg): Viral protein found in the high replicative phase of HBV. HBeAg is usually a marker of high levels of replication with wild-type viruses but is not essential for viral replication.

HB surface Antibody (anti-HBs): Antibody to HBsAg. Develops in response to hepatitis B vaccination and during recovery from hepatitis B, denoting past infection and immunity.

HB core Antibody (anti-HBc): Antibody to HBV core (capsid) protein. Anti-HBc antibodies are non-neutralizing antibodies detected in acute and chronic infections. This is especially significant during the window period.

anti-HBc IgM: Subclass of anti-HBc. Detected in recent HBV infections but can be seen by sensitive assays in chronic HBV infections.

HBV e Antibody (anti-HBe): Antibody to HBeAg. Detected in persons with lower levels of HBV replication but also in HBeAg-negative disease (*i.e.*, HBV that does not express HBeAg).

HBV DNA: HBV viral genomes that can be detected and quantified in serum by nucleic acid testing (NAT).

Serological Assays: Assays that detect the presence of either antigens or antibodies, typically in serum or plasma but also in capillary/venous whole blood and oral fluid. These include rapid diagnostic tests (RDTs) and laboratory-based immunoassays, *e.g.*, enzyme immunoassays (EIAs), chemiluminescence immunoassays (CLIAs), and electrochemiluminescence immunoassays (ECLs).

REVIEW QUESTIONS

1. What molecular aspects explain an intermediate mutation rate between a DNA virus and one of the RNAs in HBV?
2. Why was the HBV vaccine designed with recombinant HBsAg?
3. Regarding the viral load, what level indicates the emergence of mutations resulting in resistance to analogs of nucleotides?

REFERENCES

[1] Degertekin B, Hussain M, Tan J, Oberhelman K, Lok AS. Sensitivity and accuracy of an updated line probe assay (HBV DR v.3) in detecting mutations associated with hepatitis B antiviral resistance. J Hepatol 2009; 50(1): 42-8. [http://dx.doi.org/10.1016/j.jhep.2008.08.020] [PMID: 19019484]

[2] Chen CJ, Yang HI, Iloeje UH. Hepatitis B virus DNA levels and outcomes in chronic hepatitis B. Hepatology 2009; 49(S5): S72-84. [http://dx.doi.org/10.1002/hep.22884] [PMID: 19399801]

[3] Fattovich G, Bortolotti F, Donato F. Natural history of chronic hepatitis B: Special emphasis on disease progression and prognostic factors. J Hepatol 2008; 48(2): 335-52. [http://dx.doi.org/10.1016/j.jhep.2007.11.011] [PMID: 18096267]

[4] Feld J, Lee JY, Locarnini S. New targets and possible new therapeutic approaches in the chemotherapy of chronic hepatitis B. Hepatology 2003; 38(3): 545-53.

[http://dx.doi.org/10.1053/jhep.2003.50389] [PMID: 12939579]

[5] Pawlotsky JM, Dusheiko G, Hatzakis A, *et al.* Virologic monitoring of hepatitis B virus therapy in clinical trials and practice: Recommendations for a standardized approach. Gastroenterology 2008; 134(2): 405-15.
[http://dx.doi.org/10.1053/j.gastro.2007.11.036] [PMID: 18242209]

[6] Rotman Y, Brown TA, Hoofnagle JH. Evaluation of the patient with hepatitis B. Hepatology 2009; 49(S5) (Suppl.): S22-7.
[http://dx.doi.org/10.1002/hep.22976] [PMID: 19399815]

CHAPTER 16

Molecular Basis of Hepatitis C

Muneeb Ur Rehman[1] and **Hafiz Zain Ul Abideen**[2,*]

[1] *Medical College of Wuhan, University of Science and Technology, Hubei, Wuhan, China*

[2] *College of Basic and Clinical Medicine, Dali University, Dali, Yunnan, China*

Abstract: Hepatitis C virus (HCV) is a significant cause of chronic liver disease worldwide. The molecular basis of HCV infection and replication has been extensively studied, leading to the identification of vital viral proteins and their interactions with host factors. The HCV genome encodes a single polyprotein cleaved by host and viral proteases into individual proteins, including the core, envelope glycoproteins (E1 and E2), p7, NS2, NS3, NS4A, NS4B, NS5A, and NS5B. These viral proteins play critical roles in virus assembly, entry, replication, and evasion of host immune responses. The HCV envelope glycoproteins E1 and E2 are responsible for virus attachment and entry into host cells through interactions with various host receptors, including CD81, scavenger receptor class B type I (SR-BI), and tight junction proteins. The viral protein NS3 has multiple functions, including protease and helicase activities, which are critical for viral RNA replication. NS5A is an essential component of the viral replication complex and regulates viral RNA replication, virion assembly, and modulation of host immune responses. NS5B is the RNA-dependent RNA polymerase responsible for viral RNA synthesis. The molecular mechanisms underlying HCV-induced pathogenesis and the development of chronic infection remain poorly understood. However, recent studies have shed light on the interactions between HCV and host factors, including the innate and adaptive immune responses and the roles of viral proteins in modulating these responses. These insights have led to new antiviral therapies, including direct-acting antivirals (DAAs) that target viral proteins in RNA replication.

Keywords: Glycoproteins, Hepatitis C virus (HCV), HCV-induced pathogenesis, Polyprotein, Virus assembly, Viral proteins, Viral RNA replication.

INTRODUCTION

The Hepatitis C virus (HCV) is a tiny RNA virus enveloped and prefers the liver, resulting in acute and chronic hepatitis in human hosts. Globally, the number of individuals impacted amounts to 200 million, as reported by the World Health

* **Corresponding author Hafiz Zain Ul Abideen:** Medical College of Wuhan, University of Science and Technology, Hubei, Wuhan, China; E-mail: muneebrx2@gmail.com

Sanaullah Sajid, Sajjad ur Rahman, Shahid Mahmood, Shayan Bashir & Mudasser Habib (Eds.)

Organization (WHO), which further states that approximately 3% of the global population suffers from chronic infection caused by this particular virus. Most infected patients remain asymptomatic for the acute phase. They cannot eliminate the virus, which causes hundreds of thousands of patients to die yearly from chronic complications such as cirrhosis and hepatocellular carcinoma (HCC). To date, no vaccine is available to prevent infection, and the most suitable to combat it is expensive. It has adverse side effects, is unsuitable for all infected subjects, and is only effective in about 50% of patients treated. One of the limitations in advancing knowledge of hepatitis C is that there is a lack of models for small animal study since this virus only infects chimpanzees and human beings. However, the development of *in vitro* culture systems is causing considerable advancement in the knowledge related to the replication of viruses and the development of new antiviral therapies [1].

GLOBAL EPIDEMIOLOGY OF HEPATITIS C

There are about 200 million people in the world affected by HCV [2]. The prevalence varies by region: it is the highest in North Africa and the Middle East, mainly Egypt, where a prevalence of 14%, followed by the rest of Africa (3%), China and other Asian countries (2.1%), Eastern Europe and the United States (1.6%). In northern European countries, such as the United Kingdom, a prevalence of 1.1% is reported. In Mexico, although studies are not available on a representative population, and according to the existing reports, most blood donors' prevalence of hepatitis C is around 0.4%.

Transmission Routes

Contact with infected blood or products is the most prevalent way that people contract HCV. The spread of HCV has been significantly impacted by injectable therapy and medicines.

Intravenous Drug Use (IVD): In developed countries, such as the United States, IVD is the leading risk factor, even though a decrease in cases has been observed. The prevalence of hepatitis C in users who had been consuming IVD for a year decreased to 10% from 1997 to 1998, while from 1988 to 1991, the reported prevalence was 65%. In countries with high prevalence, most infections are acquired by not having safe blood transfusion measures, administering injections with non-disposable syringes, and using inefficient cutting material in medical and dental clinics.

Sexual Contact: Of cases with acute hepatitis reported in the United States, 15 to 20% of patients are engaged in high-risk sexual activity (more than two sexual partners during six months or having relationships with infected people).

However, prevalence studies conducted on monogamous couples where one is infected with HCV suggest that the spread of hepatitis C by sexual contact is rare.

Perinatal Transmission: Transmission to newborns, the number of infected mothers varies from 4.6 to 10%. The only element causing the transmission of hepatitis C in newborns is the presence of viral RNA in mothers during childbirth.

Transmission to Health Personnel: In developed countries, it is rare to observe patient-to-patient messages. When detected, this is related to improper healing practices, such as using multi-dose vials, improper cleaning, lack of equipment, and lack of aseptic techniques. The transmission of HCV from infected patients to health personnel is mainly due to accidents with contaminated needles.

Structural Features of HCV

HCV belongs to the Flaviviridae family and is the only member of the Hepaciviral genus. There are six genotypes (1 to 6) identified according to their nucleic acid sequence, with a divergence in its complete genome of 30 to 50%; there are also more than 70 sub-genotypes (a, b, c, d, *etc.*), identified by differences in the comparison of the genome from 20 to 25% among the genotypes. Quasispecies refers to the mixture of HCV genetic variants in the same individual. Of the identified genotypes, the most resistant to treatment is genotype 1. HCV is a single-stranded RNA virus, with sense positive, replicated by an RNA-dependent RNA polymerase. Its main organ of replication is the liver; however, evidence is available of extrahepatic cells for HCV, such as B lymphocytes from peripheral blood, epithelial cells in the intestine, and the central nervous system. The viral particles, analyzed in the electron microscope, are observed heterogeneously: have at least two sizes (50 and 100 nm in diameter) and contain double-walled and beak-shaped projections on their surface [3].

As was previously mentioned, the HCV genome consists of a 9.6 kb positive-sense single-stranded RNA. It has a single open reading frame (MLA) that codes for a polypeptide precursor of about 3000 amino acids in length. At least ten distinct proteins are produced from this polyprotein after it is cleaved by viral proteases and the host (Fig. **1**).

- A vital membrane protein is called p7; three structural proteins are called core, E1, and E2.
- The intracellular steps of the HCV life cycle are coordinated by six non-structural proteins: NS2, NS3, NS4A, NS4B, NS5A, and NS5B.

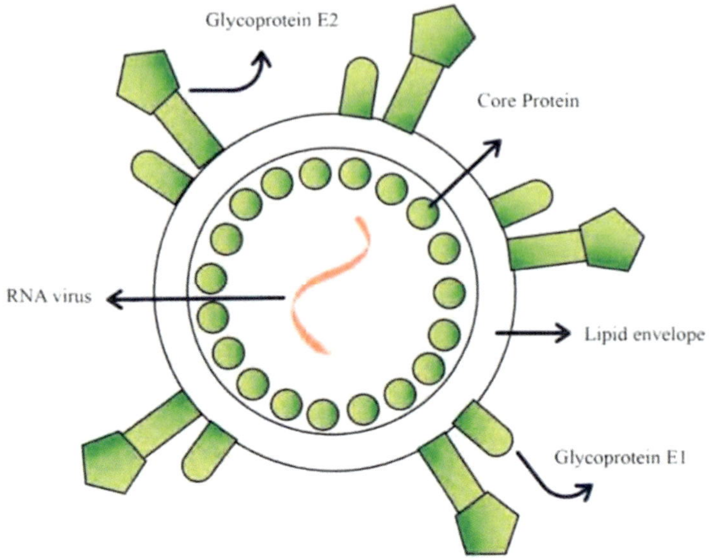

Fig. (1). A schematic with the structural components of the HCV particle

Hepatitis C Virus, its Proteins, and Genes

HCV RNA lacks the 5'cap site, and its translation depends on an internal ribosome entry site (IRES) located in the region 5'-non-encode sing (5' NCR). The IRES is attached to the 40S subunit of the ribosome. Once the HCV genome translation has started, a long polyprotein is produced by the protease-mediated proteolytic attack of the host and the virus, and the structural proteins are obtained. The organization of the viral genome has the same structure as the polyprotein, with the non-coding regions located at the 5' and 3' ends (5'-RNC and 3'-RNC, respectively) of the viral genome (Fig. **2**).

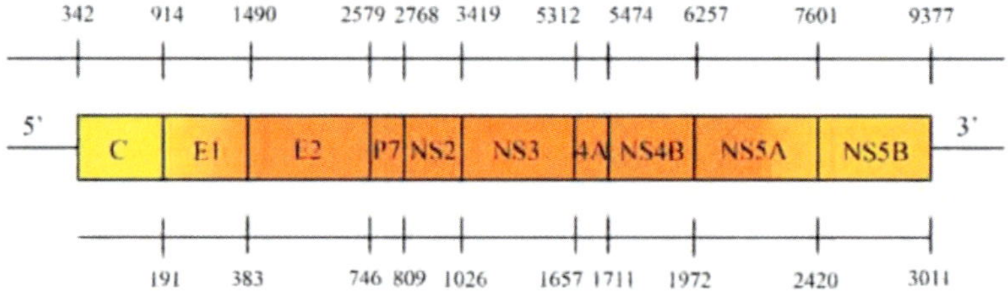

Fig. (2). Linear organization of the genome of hepatitis C. HCV encodes at least ten proteins per second of an open reading frame, giving rise to a polyprotein precursor cleaved into various structural and non-structural proteins. The upper line specifies the position of the ribonucleotide, and on the line below is the amino acid, which results in the initiation of each viral protein.

Structural Proteins

Core Protein

The HCV core protein contains a high proportion of essential amino acids. This feature allows it to bind to the viral RNA, forming a nucleocapsid.

Glycoproteins Envelope

The glycoproteins envelope E1 and E2 are type I transmembrane proteins that form heterodimers that are covalent and include the building blocks of the viral envelope participating in the entry of the virus into the cell.

P7

It is a polypeptide of 63 amino acids; within the polyprotein, the viral envelope is localized in the binding part of the structural and non-structural proteins. P7 contains two transmembrane domains that form hexamers with ion channel activity. It is unknown if P7 assembles into the viral particles.

Non-Structural Proteins

NS2/3 Auto Protease

In the viral polyprotein, the binding of NS2/3 is hydrolyzed utilizing its auto-protease activity, which is localized in the NH-terminal of NS3; this process is essential for the replication of the virus. The NS2 protein derived from the break of NS2/3 is inserted into the endoplasmic reticulum membrane and participates in the phosphorylation of the viral protein NS5A, an essential event for viral replication.

NS4A and NS4B

NS4A is a 54 amino acid polypeptide cofactor of the serine protease located in the NS3 domain. NS4B is a highly hydrophobic 27 kDa protein that integrates into the endoplasmic reticulum's membrane. This protein induces an alteration in the membrane, known as a membranous network, which serves as a site for the formation the viral replication complex.

NS5A and NS5B

Zinc is one of the metals in NS5A, which is a phosphorylated metalloprotein involved in viral replication and particle assembly. The protein's crystal structure is thought to facilitate the assembly of the replication complex from viral RNA.

New RNA genomes are synthesized with the help of NS5B, the viral polymerase, an RNA-dependent RNA polymerase. The intermediate negative sense complementary RNA strand is synthesized from the positive strand of viral RNA during replication. This strand is then employed as a template to synthesize additional positive-stranded viral RNAs.

HEPATITIS C VIRUS REPLICATION

Similar to other positive-stranded RNA viruses, Hepatitis C Virus (HCV) establishes a replication complex that is connected with cellular membranes. This complex consists of viral proteins, viral RNA, modified cell membranes, and components derived from the host organism. The glycoprotein E2 exhibits a robust binding affinity towards CD81, a protein that is ubiquitously present on the cellular membrane of various cell types, including hepatocytes. However, in addition to E2 and CD81, more molecules are required to enter HCV into the cell, including the class B type I receptor (SRB1), which regulates cholesterol metabolism. This receptor, in conjunction with high-density lipoproteins (HDL), plays a significant role in the onset of infection for HCV. The viral protein E1 is considered to participate in the fusion of the virus with the membrane on the cytoplasmic face. The low-density lipoprotein receptor (LDL), glycosaminoglycans, and heparan sulfate enter HCV into the hepatocyte. Once HCV enters the cell, the nucleocapsid is released into the cytoplasm, where the viral RNA functions as a template in two critical events of the replication: 1) the polyprotein translation and 2) the replication of the Viral RNA.

During translation, the polyprotein is produced as a precursor, which is processed into structural (C, E1, E2, P7) and non-structural (NS2, NS3, NS4A, NS4B, NS5A, and NS5B) proteins by the proteolytic attack of the proteases of the viral host. The viral RNA-positive strand replication occurs in a membranous compartment in the cell's cytoplasm. It gives rise to a negative-stranded RNA, forming double-stranded RNA, an intermediary for the new formation of positive-stranded viral RNA. These processes allow the assembly of the structural proteins of the virus with the Viral RNA to form new viral particles, which are released from the cell by exocytosis.

Natural History of Hepatitis

Acute Hepatitis C

Most patients infected with HCV do not present a picture of acute hepatitis. The acute infection of HCV is rarely symptomatic (15%). During the evolution from acute hepatitis C, four phases can be distinguished: the first is characterized by the non-detection or detection of very low circulating HCV RNA concentrations;

about two weeks later, it follows a short phase of eight to 10 days, in which there is a rapid increase in circulating viral load; the third phase is characterized by a stationary stage in the circulating levels of the viral RNA, with a duration of about 40 to 60 days; finally, the fourth phase is characterized by the detection of anti-HCV antibodies, a decrease in viral load or progression to a chronic disease state in most infected patients. The presence of antibodies is usually detected in the ninth week; however, there are cases in which it has been reported that it takes up to a year to appear after the first detection of viremia, and approximately 15 to 40% of patients exposed to HCV apparent the infection in six months [4] (Fig. **3**).

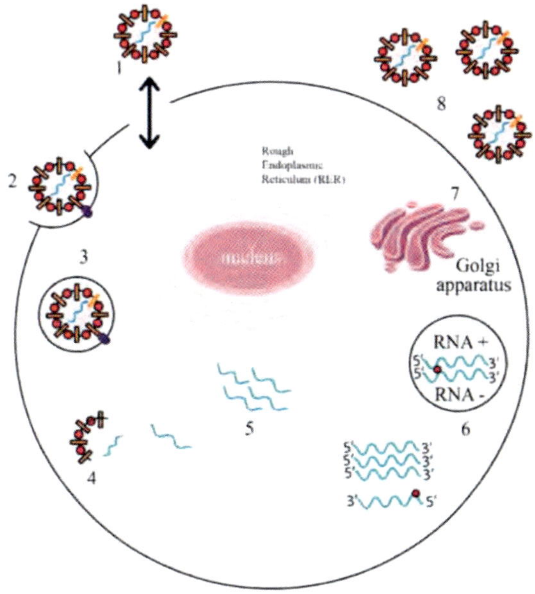

Fig. (3). HCV life cycle. 1) Binding of viral glycoproteins with cell membrane receptors. 2) Fusion of membrane. 3) Viral entry into the cell. 4) Loss of the viral nucleocapsid. 5) Synthesis of viral proteins in the rough endoplasmic reticulum (RER). 6) Viral replication in the cytoplasm. 7) Encapsulation of viral RNA membranes in the Golgi apparatus. 8) Release of viruses.

Chronic Hepatitis C

Most people infected with HCV develop chronicity. Studies conducted on blood donors, people who have received blood products, and IVD users from 55 to 86% of people infected with HCV evolve to chronicity. In chronic patients, the course of the disease is variable, and not everyone develops liver complications such as cirrhosis or HCC. Studies conducted on liver clinic patients show that 18 to 26% of these subjects have liver complications, such as cirrhosis. Of patients who develop cirrhosis, the likelihood of presenting decompensated liver disease is 4% per year. In patients with advanced fibrosis, 7% evolve to HCC every year [4].

MOLECULAR DIAGNOSIS OF HEPATITIS C

HCV infection is often underdiagnosed, and up to 50% of people with infection by this virus are not aware of their illness. Therefore, molecular techniques are used as a critical tool in the diagnosis and monitoring of treatment for infection with HCV. Viral RNA can be detected in serum, liver, and peripheral blood mononuclear cells of patients infected with acute, chronic hepatitis, cirrhosis, and HCC. Detecting viral RNA in serum is considered the most sensitive and specific indicator of viral replication and infectivity for HCV. Confirming the serological diagnosis and the patient's follow-up treatment is essential. The tests used to determine the presence of viral RNA are classified into qualitative and quantitative and are very sensitive and specific [5].

Qualitative tests document the presence or absence of the virus in the blood; of these tests, the most commonly used is the reverse transcription (RT) followed by the reaction in a polymerase chain reaction (RT-PCR). This test is available in various commercial kits; generally, the sensitivity is 96%, the specificity is 99%, and they have a sensitive detection limit. In addition, to confirm viremia, these tests are beneficial in the search for HCV in blood product donors, which increases the prevention and decreases the contagion of this virus since, in most of the world, they use the same antibodies; on the other hand, they have the disadvantage of presenting a broad window period. Quantitative tests document the number of viral RNA copies in international units per milliliter (IU/ml) (viral load). The conversion of the UI to copies/ml varies depending on the commercial equipment being used but ranges from 0.9 to 5.2 copies/ml. Several tests are available quantitatively, based on real-time PCR, which can accurately quantify the levels of viral RNA in a linear range exceeding six logarithms.

DETERMINATION OF THE VIRAL GENOTYPE

As mentioned above, HCV is classified into six genotypes and several subgenotypes. It is clinically essential to carry out the molecular diagnosis of these genotypes since they predict the probability of response to treatment, dictate its duration, and often determine the drug's dosage. The genotype is the stronger predictor of response to interferon and ribavirin. For the determination of the viral genotype, there are two commercial methods: one is based on the sequencing technique of nucleic acids in the 5' NC region of the virus (True Gene 5'NC HCV genotyping kit), and an RT-PCR carries out the other, followed by a reverse hybridization, using genotype-specific probes that bind in the 5'NC region (Versant HCV genotype Assay LiPA). Once the viral load is known, the viral genotype and all clinical and laboratory factors that evaluate the effect of HCV on the deterioration of liver function can establish the treatment criteria in the patient.

TREATMENT OF HEPATITIS C

Since most HCV-infected patients remain asymptomatic during the acute phase, those who come to receive the treatment are predominantly patients with chronic liver damage. Therapy aims to achieve a sustained virological response (SVR), that is, elimination of the viral load for at least six months after the conclusion of treatment. No selective antiviral therapy is available, not even a preventive vaccine. The initial treatment for hepatitis C consisted of monotherapy with interferon α (IFN-α). The results showed considerable improvement when the IFN-α molecule was chemically modified using polyethylene glycol, known as pegylated IFN (PEG-IFN). This modification increased the compound's half-life, allowing it to remain in blood circulation for days and be active. This improvement in the bioavailability of IFN *in vivo* and the addition of ribavirin to treatment improved more than 100% in SVR compared to non-pegylated IFN-α. Even with these changes, which improved the treatment for hepatitis C, it is essential to note that only about 50% of the patients treated get an SRV. The HCV genotype plays a vital role in the response and speed of an SRV to treatment. About 50% of patients with genotype 1 develop a favorable SVR compared to more than 80% of those with genotype 2 or 3 who respond well.

- The oral medications currently employed in the treatment of hepatitis C include Glecaprevir, Pibrentasvir, Sofosbuvir, Velpatasvir, Voxilaprevir, Elbasvir, and Grazoprevir. These drugs belong to different classes, such as NS3/4A protease inhibitors, NS5A inhibitors, and nucleotide analog NS5B polymerase inhibitors.
- Recent advancements in the treatment of hepatitis C include the introduction of the combination drug Mavyret (glecaprevir and pibrentasvir), the triple-drug cocktail Vosevi (sofosbuvir, velpatasvir, and voxilaprevir), and the double-drug combination Epclusa (sofosbuvir and velpatasvir).

  The FDA has approved Epclusa, Mavyret, and Vosevi to treat chronic hepatitis C virus (HCV) genotype 1, 2, 3, 4, 5, or 6 infections in people who do not have cirrhosis or who have cirrhosis that has been fixed (Child-Pugh A).
- Individuals at least three years of age are eligible for therapy with Mavyret. Mavyret is licensed to treat patients aged three years and above licensure for treating patients aged three years and above who are infected with Hepatitis C Virus (HCV) genotype one. These patients must have previously had treatment, including either an HCV NS5A inhibitor or an NS3/4A protease inhibitor (PI), but not both. These patients have not had treatment involving administering both types of inhibitors.
- One of the indications for the usage of Vosevi is the treatment of adult patients. Eligibility for re-treatment with this medicine is extended to patients who have previously had therapy with a regimen incorporating either an NS5A inhibitor

(GT1-6) or a regimen using sofosbuvir but without an NS5A inhibitor (GT1a or GT3).

- Epclusa has been approved for use in people aged three and up, making it available to adults and children. Epclusa (combined with ribavirin) is also approved for treating patients with decompensated cirrhosis with HCV genotypes 1, 2, 3, 4, 5, or 6. Anyone over three, including adults, can receive this treatment.

The best treatment has not yet been established for acute HCV infection. Still, several studies have observed good results using 5 MIU of IFN daily for four weeks, followed by 5 MIU three times a week for 20 weeks; this scheme has demonstrated SVR in more than 95% of patients. The most important thing about this is that, during acute infection by HCV, the genotype and serum RNA levels appear not to influence treatment. For the treatment of chronic HCV infection, guidelines recommend PEG-IFN plus 1000 to 1200 mg ribavirin for 48 weeks for genotypes 1 or 4, and PEG-IFN plus 800 mg ribavirin for 24 weeks for genotype 2 or 3, with which it has shown SVR of 55 and 80%, respectively. Those who do not respond to treatment and continue with chronic liver disease can evolve with severe consequences, such as developing HCC or presenting a liver decompensation. The option for these patients could be a liver transplant; however, it has been observed that if they offer an active infection by HCV during transplantation, there is an increase of 10-fold in viral load relative to the levels shown before the procedure. Of those who develop cirrhosis after transplantation, about 40% will present liver decompensation, and only 50% of these patients will survive a year after decompensation. New drugs are under development, explicitly inhibiting a site in the life cycle of HCV. In addition, biological, molecular, and recombinant DNA technology has reached the production of molecules that could interfere with the synthesis of the virus, such as ribozymes and interfering DNA; however, this is still under *in vitro* research.

GLOSSARY

Anti-HCV Antibody: The presence of antibodies to the Hepatitis C virus (HCV) may typically be identified in the bloodstream around two to three months following either HCV infection or exposure.

HCV RNA: Hepatitis C virus (HCV) viral genomes can be identified and measured in serum through nucleic acid testing (NAT).

HCV core antigen (HCVcAg): Released into plasma during viral assembly, HCV nucleocapsid peptide 22 [p22] is detectable at all stages of infection.

HCV Sustained Virological Response (SVR): HCV RNA levels in the blood are undetectable at a predetermined time post-treatment, typically 12 or 24 weeks (SVR12 or 24).

HCV Relapse: HCV RNA is undetectable during therapy and at treatment's completion but is detected after treatment's conclusion.

HCV Viral Breakthrough: HCV RNA is undetectable throughout treatment but after that present despite ongoing treatment.

REVIEW QUESTIONS

1. Which HCV protein interacts with the CD81 receptor to enter the hepatocyte?
2. How is it determined that hepatitis C evolved to chronicity?
3. In patients with chronic hepatitis, what are the most resistant genotypes of HCV to develop a sustained virological response?

REFERENCES

[1] Alter MJ. Epidemiology of hepatitis C virus infection. World J Gastroenterol 2007; 13(17): 2436-41. [http://dx.doi.org/10.3748/wjg.v13.i17.2436] [PMID: 17552026]

[2] Asselah T, Bieche I, Paradis V, Bedossa P, Vidaud M, Marcellin P. Genetics, genomics, and proteomics: implications for the diagnosis and the treatment of chronic hepatitis C InSeminars in liver disease . Inc., 333 Seventh Avenue, New York, NY 10001, USA.: Copyright© 2007 by Thieme Medical Publishers 2007.

[3] Irving WL, Brown RJP. Acute hepatitis C virus infection: A dynamic-and challenging-concept. J Infect Dis 2010; 202(12): 1765-7. [http://dx.doi.org/10.1086/657318] [PMID: 21067368]

[4] Narahari S, Juwle A, Basak S, Saranath D. Prevalence and geographic distribution of Hepatitis C Virus genotypes in Indian patient cohort. Infect Genet Evol 2009; 9(4): 643-5. [http://dx.doi.org/10.1016/j.meegid.2009.04.001] [PMID: 19460332]

[5] Pawlotsky JM, Chevaliez S, McHutchison JG. The hepatitis C virus life cycle as a target for new antiviral therapies. Gastroenterology 2007; 132(5): 1979-98. [http://dx.doi.org/10.1053/j.gastro.2007.03.116] [PMID: 17484890]

CHAPTER 17

Molecular Basis of the Human Immunodeficiency Virus

Maliha Sarfraz[1,*], **Sanaullah Sajid**[2] and **Hayat Ullah**[3]

[1] *Department of Zoology, Wildlife and Fisheries, University of Agriculture Faisalabad, Sub-Campus, Toba Tek Singh, Pakistan*

[2] *Institute of Microbiology, University of Agriculture, Faisalabad, Pakistan*

[3] *Department of Chemistry, University of Okara, Okara, Punjab, 56300, Pakistan*

Abstract: Human Immunodeficiency Virus (HIV) is the causative agent of Acquired Immunodeficiency Syndrome (AIDS), a deadly disease that affects the human immune system. HIV is a retrovirus that infects T-cells, macrophages, and dendritic cells in the immune system, leading to their destruction and, ultimately, the onset of AIDS. The molecular basis of HIV infection involves the interaction of the viral envelope protein gp120 with the host cell receptor CD4 and a co-receptor such as CCR5 or CXCR4. This binding triggers a conformational change in gp120 that exposes a fusion peptide, allowing the viral envelope to fuse with the host cell membrane and release its contents into the cytoplasm. Once inside the host cell, the viral genome is reverse-transcribed into DNA, which is then integrated into the host cell genome by the viral integrase enzyme. This allows the virus to replicate with the host cell and evade the immune system's surveillance. Despite advances in antiretroviral therapy, HIV continues to pose a significant global health threat, with over 38 million people living with HIV/AIDS worldwide. Understanding the molecular mechanisms of HIV infection is critical for developing effective treatments and vaccines to combat this deadly disease.

Keywords: AIDS, Antiretroviral therapy, CD4, Co-receptor, Dendritic cells, HIV, Macrophages, Retrovirus, Reverse transcription, T-cells, Viral envelope protein, Viral genome.

INTRODUCTION

Viruses are classified into more than 60 families, depending on the type of nucleic acid they contain, the information in their genome, and their type of genes. Among these families, at least 20 infect humans. The causative agent of acquired immunodeficiency syndrome (AIDS) is a retrovirus known as the human immu-

* **Corresponding author Maliha Sarfraz:** Department of Zoology, Wildlife and Fisheries, University of Agriculture Faisalabad, Sub-Campus, Toba Tek Singh, Pakistan; E-mail: maliha.sarfraz@uaf.edu.pk

Sanaullah Sajid, Sajjad ur Rahman, Shahid Mahmood, Shayan Bashir & Mudasser Habib (Eds.)

nodeficiency virus (HIV), which belongs to the Retroviridae family and the Lentivirus genus. HIV was first identified in France in 1983 by Luc Montagnier, the virus with an RNA genome that induces chronic cell infection by converting its RNA to proviral DNA, which integrates into the cell's genome and induces progressive damage to the immune system of the host. Infection by this lentivirus is characterized by a slow progression, with long periods of clinical latency, followed by the gradual onset of signs and symptoms, such as fever, lymphadenopathy, opportunistic diseases, and oncological and chronic degenerative diseases.

One of the main characteristics of HIV is its genetic variability, which shows during its evolution the transmission of simian virus genomic fragments to the human that has given rise to various types, subtypes, variants, and multiple recombinant forms. Another factor influencing genetic variability is the virus genome's high mutation rate, which is provided by the reverse transcriptase enzyme. This enzyme generates 3×10^5 mutations per nucleotide per cycle and a production of 1010 viral particles per day. Considering these two variables, it is estimated that, on average, 3×10^9 can be generated mutations per day in the viral population of each person with HIV. Most of the mutations generated in the variants have little or no effect on viral function or replication capacity; still, others can be lethal for the virus. However, when these mutations are located at key genomic sites, such as genes for viral enzymes, they affect the virus's ability or may confer resistance to one or more antiretrovirals. Another factor contributing to genetic diversity is the recombination of the two RNA strands, which constitutes an intrinsic part of the normal cycle of viral replication and can mediate the repair of defective viral genomes, increase viral diversity or accelerate the spread of mutations beneficial among viral quasispecies. Genetic variability impacts the pandemic of HIV through diagnosis, pathogenesis, progression, and transmission of the disease, the quantifiability of viral load, clinical management, response to antiretroviral treatment, and vaccine design [1].

NATURAL HISTORY OF DISEASE

The various clinical states of HIV infection result from changes in the balance of immunological and virological status. Immune status is determined by the count of TCD4+ lymphocytes and is expressed as cells/mm^3. Virologic status is determined by the copy number of HIV RNA (viral load) and expressed as copies/ml. These are characteristics of the infection and the gradual destruction of the CD4+ T lymphocyte population with constant production of new viral particles, which generates immunodeficiency and takes the patient to the last stage of HIV infection or AIDS. HIV-infected patients go through three clinical stages [2].

1. Primary infection or acute phase: seroconversion.
2. Asymptomatic state or clinical latency.
3. Symptomatic state or AIDS.

Acute Phase

This period begins with the patient's infection and may have no clinical manifestation since it is only symptomatic in approximately 40-80% of patients two to three weeks after infection. The symptoms can vary, including fever, mononucleosis, maculopapular rash, sore throat, general malaise, and joint pain. Signs and symptoms disappear in two or three weeks; later, the patient is asymptomatic between six and ten years. The first episodes that occur after HIV crosses the mucosal barrier are known as the window period, during which the production of IgM and IgG antibodies is not sufficient for the diagnostic tests to detect the immune response against HIV. The duration is controversial, although historically, it has been suggested that it takes approximately three months. Due to the nonspecific nature of symptoms during this stage, the diagnosis of HIV disease is rarely recognized in medical practice. Between two and six weeks after infection, the viral load in the plasma reaches hundreds of thousands of copies/ml. Due to the massive increase in viral replication, the number of CD4+ T cells decreases. Consequently, the elevated viral load also contributes to the transmission of the disease. It has been estimated that up to 30% of infections by HIV are transmitted during the acute phase.

Four to 12 weeks after infection, specific IgG antibodies appear against the HIV, and the specific cellular immune response is activated, the leading cause of the decrease in viral load and the corresponding increase in the CD4+ T lymphocyte count in patients. A balance is maintained between the number of viral particles produced and eliminated and the number of lymphocytes T CD4+ destroyed and generated each day six to twelve months after infection.

Clinical Latency Phase

Despite the absence of early symptoms, the majority of patients are diagnosed at this stage. During this stage, high viral replication persists, and the number of viral particles produced reaches more than 1010 particles per day. The ability to counteract these particles to the regeneration of CD4+ T lymphocytes, however, is not sufficient to halt the progression of the illness. With the appropriate use of antiretroviral treatment, which inhibits the replication of the virus in different stages of the viral replication cycle, the reconstitution of the immune system is possible, which prolongs the appearance of the symptoms of AIDS and increases the quantity and quality of life of people living with HIV. The duration of this asymptomatic, also known as chronic, phase is variable and depends on a variety of factors, including both the virus and the infected individual. Approximately 80 to 90% of patients with this stage live for 8 to 10 years.

SYMPTOMATIC STAGE OR AIDS

The continued wear and tear on the immune system due to the loss of CD4+ T lymphocytes increase the susceptibility to various infections. At this stage, there is an increased viral replication activity, and the immune system can no longer stop it. Clinically, when the CD4+ lymphocyte count drops to <200 cells/mm^3, patients are more susceptible to opportunistic infections that can lead to death. Neoplasms such as Kaposi's sarcoma, tuberculosis, pneumonia, candidiasis, and certain neurological disorders are some characteristic infections of the AIDS stage. The duration of this phase is unfavorable, with a survival of less than 15 to 30% at three years. In this disease state, the immune system is highly compromised; even if the patient receives antiretroviral treatment, he will eventually die.

CLASSIFICATION OF HIV

The molecular epidemiology of HIV is a complex subject constantly evolving due to the high genetic variability inherent to this retrovirus. HIV emerged in West and Central Africa as a result of the transmission of multiple zoonoses of immunodeficiency viruses from apes and nonhuman primates to humans. Several hypotheses explain the transmission between primates and humans; the most accepted are: a) the process of hunting and killing primates for meat consumption, b) the direct exposure of humans to the blood or mucous secretions of animals, and c) the contamination in the preparation of the oral vaccine against polio.

The classification of HIV is elaborated according to the patterns grouped by genetic similarity. To date, two types of HIV have been identified: HIV-1 and HIV-2 [3].

HIV-1

Through numerous phylogenetic analyses of the HIV strains isolated in various geographical areas of the world, the virus has been classified into four categories: groups, subtypes, and recombinant circulating forms. There are four groups of HIV-1, transmitted independently from chimpanzee and gorilla to human, labeled M, N, O, and P. Groups M and N have a genetic relationship to the viruses found in chimpanzee, while the groups O and P found in gorillas. The groups are called “M” for primary or principal, “O” for the outlier, and “N” for non-M, non-O. The nomenclature to classify HIV strains was proposed in 1999; it was agreed that the new groups would be assigned a letter in strict alphabetical order. Group M includes the strains responsible for the pandemic worldwide and is subdivided into nine subtypes. (A-D, F-H, J, and K). The A and F subtypes are further divided into sub-subtypes (A1, A2, F1, and F2). As new subtypes are found, they will also be named in alphabetical order. Some strains are classified as having different subtypes in different regions of their genomes based on phylogenetic analysis. The mosaic genomes are referred to as circulating recombinant forms (CRFs).

At present, 52 CRFs have been described that are identified with the numbers and letters of the subtypes that are involved. When they recombine more than three subtypes, they are called cpx (complex) and are classified in the order in which they are described. In the world, subtype C represents 48% of infections, followed by subtypes A, B, G, and D, which account for 12, 11, 5, and 2% of infections, respectively. The CRFs account for at least 20% of HIV infections worldwide. Group N is formed by closely related phylogenetic sequences since it has only been described in a few individuals from Cameroon. Group O was initially found in West-Central Africa, and, like group M, its sequence diversity is high. It is classified into I-V classes, which are genetically distant from each other like the M group subtypes; however, they have not spread to everyone. Group P was recently described, and only a few sequences are available in individuals originating from Cameroon.

HIV-2

The HIV-2 virus in Sooty mangabey apes is closely related to the HIV-2 virus in humans. It is divided into groups from A to H. This virus is endemic to West Africa, and phylogenetic studies show that the various groups of this virus originated from independent transmissions between the Sooty mangabey and humans.

VIRAL GENOME

The HIV genome comprises approximately 10 kb and nine genes: three structural, characteristics of retroviruses; two regulators; and four accessory genes. In addition, at the ends, it has long terminal repeats (LTR), which function as promoter regions for the transcription of the virus. To synthesize new viral particles, structural proteins are required, and the genes that encode these proteins are Gag, Pol, and Env. The Gag gene codes for the capsid proteins (p24, p17, p9, and p7); p24 forms the nucleocapsid; p17 forms the virus matrix between the virus's outer envelope and the nucleocapsid, and p7 and p9 are closely associated with viral RNA. The Pol gene codes for the three essential enzymes of the viral replication cycle: the reverse transcriptase enzyme (p66)/RNase H, integrase (p31), and protease. The Env gene codes for the envelope precursor protein and is processed by a cellular protease into two envelope proteins: gp120, found on the virus's surface, and gp41, a transmembrane protein; this glycoprotein complex that binds to the cellular receptor.

The regulatory genes are Tat and Rev, whose proteins are essential for viral replication within the cell. The function of the Tat protein is to regulate transcription from the viral promoter, and that of the Rev protein assists in the efficient transport of viral RNA from the nucleus to the cytoplasm; in the absence of Rev, structural proteins are not formed. Accessory genes are involved in regulating viral gene expression and cell latency. These genes are nef, vif, vpr, vpu, and vpx. The genes nef, vif, and vpr are found in all retroviruses except SIVagm, which does not contain vpr. HIV-1 contains vpu, whereas HIV-2 contains vpx. The nef gene codes for a protein required for efficient viral replication interacts with host cell components, and is associated with high viral loads. The Nef protein has at least two distinct functions: it improves the replication of the virus. It stimulates a reduction in the number of CD4 receptor cells on the infected cell's surface, which is used to prevent reinfections. Another biological function of nef is regulating the MHC-1 molecule in the infected cells, which masks the immune system.

The Vif protein has an effect on free and infectious viral particles. The Vpr protein serves several purposes during viral replication in cells: it affects the transcription process, the importation of the viral DNA preintegration complex, and the regulation of apoptosis, and it intervenes in the host cell's genes. Vpu protein, found only in HIV-1, facilitates the release of viral particles, intervenes in the endoplasmic reticulum for the expression of the CD4 cell receptor, and is involved in the maturation of envelope proteins; without this protein, the viral particles remain attached to the cell membrane. Finally, the vpx gene is unique to HIV-2 and has functions similar to the Vpr protein of HIV-1, but without stopping the cell cycle. This protein is associated with the pathogenesis of the virus because, in its absence, AIDS develops.

STRUCTURE OF THE VIRION

HIV belongs to a group of isometric viruses and measures between 90 and 140 nm in diameter. Like other retroviruses, the HIV viral particle comprises the capsid protein (p24), that of the nucleocapsid (p7/p9), two strands of single-stranded RNA, and three enzymes: protease, reverse transcriptase, and integrase. The virus envelope comprises lipids derived from the host cell and viral spikes encoded by the virus. It has about 70 spicules on the lipid envelope, consisting of a transmembrane molecule, glycoprotein 41, that acts non-covalently with gp120 (outer glycoprotein). The virion envelope contains the matrix protein (p17), which delimits the outer membrane of the internal capsid. The mature HIV particle contains an internal capsid ribonucleoprotein protein comprising p24 and p7/p6. Inside the capsid are enzymes such as reverse transcriptase/RNase H, integrase, and protease; it also contains two copies of single-stranded RNA joined together through noncovalent bonds (Fig. **1**).

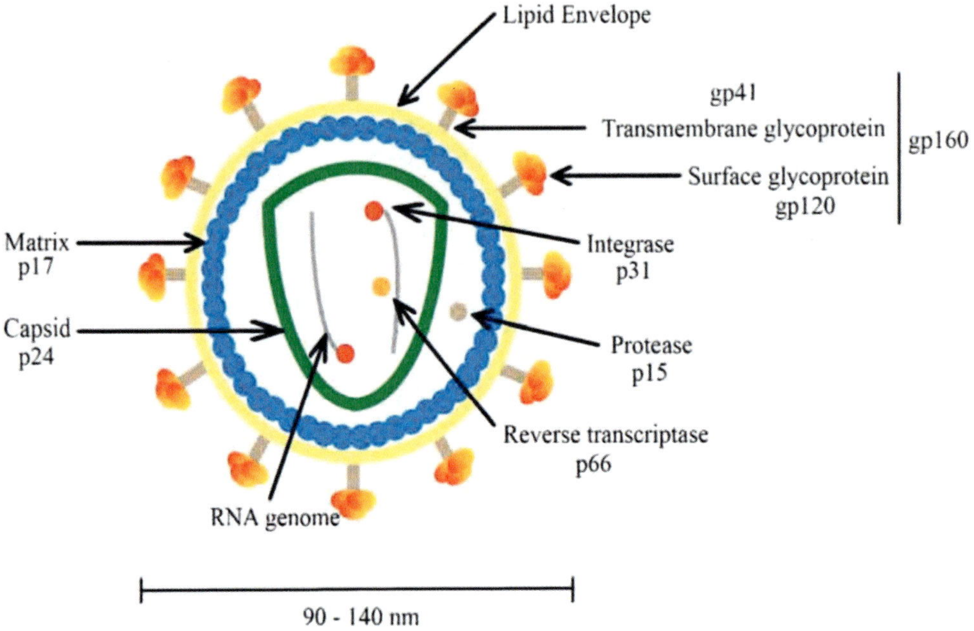

Fig. (1). Structure of HIV. The nucleocapsid contains two copies of the viral genome, core proteins, and enzymes. The surrounding envelope comprises lipids derived from the host cell and viral glycoproteins (gp120 and gp41).

HIV REPLICATION CYCLE

Like other retroviruses, HIV must replicate within a target cell using its cellular machinery. The virus initiates the infection cycle by binding the viral particle cells with the CD4+ receptor. Cells susceptible to infection are helper T lymphocytes, macrophages, Langerhans, dendritic cells, microglia, and megakaryocytes. The viral particle adheres to the cell through the interaction of the gp120 glycoprotein of the virus envelope with the CD4+ cell surface receptor. This union causes conformational changes in both receptors, allowing them to come closer together and leading to the virus's second binding with the cellular co-receptor (CCR5 or CXCR4). This second interaction generates a new conformational change in gp41, which causes a fusion pore. This union causes changes in the cell membrane and virus envelope, allowing them to fuse.

Consequently, if the capsid is partially degraded, the contents of the capsid are released, and the viral particle (among other components, the two strands of RNA and the reverse transcriptase enzyme) is in the cytoplasm of the cell. The next step in replication involves reverse transcription, where the viral RNA is transcribed into proviral DNA. The reverse transcriptase enzyme is responsible for the synthesis of the first strand of DNA (ssDNA) complementary to the RNA strand. During the synthesis of this strand, reverse transcriptase often switches from one RNA strand to the other, generating a single-strand DNA with the recombination of the two original strands of viral RNA. The Rnase-H activity of the enzyme degrades the ARNA when it has already been copied.

Subsequently, the ssDNA serves as a template to synthesize the second strand of DNA. It forms the preintegration complex that is transported *via* the endoplasmic reticulum, and with the help of the viral protein Vpr, to the nucleus cell, where it will be integrated into the host cell genome. The preintegration complex crosses the nuclear membrane and, in collaboration with host repair enzymes, it integrates the virus genome into the host cell chromosome, resulting in the proviral genome. Transcription of the proviral genome is regulated by the Tat and Rev genes, where the Tat protein helps elongate RNA and Rev regulates RNA transport from the cytoplasmic nucleus. RNA binds to ribosomes and initiates the synthesis of viral proteins, which are transported to assemble near the cell membrane. The structural proteins of the Gag, Pol, and Env genes are expressed as polyproteins, which are cleaved by the protease of the virus until they become small functional units.

The Env gene products are transported to the cell membrane, and RNA packaging requires Gag gene signals. The viral envelope proteins bind within the host cell membrane and the proteins that form the virus's core, the two RNA strands and enzymes. Finally, the release of the new viral particles involved the vpu gene. The virus pricks the cell and releases new viruses outside the cell, where the virus continues to mature with the help of protease. A single cell can produce thousands of infectious HIV particles for weeks or in a single burst after the death of the infected cell.

SEROLOGICAL AND MOLECULAR DIAGNOSIS

After exposure to the virus, generally from the second to the third week, the virological and serological events show a high viremia, which causes the distribution of the virus throughout the body. HIV RNA is present before the patient develops detectable anti-HIV antibodies. The virus's central capsid protein, the antigen p24 (Ag p24), can usually be detected within a few days of the onset of viremia. When the host's immune system initiates a response to control the infection, the levels of both virus and Ag p24 decrease to be undetectable. In general, Ag p24 will remain undetectable until the immune system gets tired and is associated with the progression of the disease. Knowing about these virological and serologic events after infection allows you to diagnose HIV infection using the techniques listed below. The diagnosis of HIV infection is based on tests from the laboratory [4].

DIAGNOSTIC METHODS

Indirect

It detects the presence of specific antibodies that the organism produces in response. This group includes screening tests, such as the enzyme-linked immunosorbent assay (ELISA) and confirmatory test (Western blot).

Direct

It detects the virus or some of its components (viral culture, polymerase chain reaction, PCR). The most frequent test is the ELISA. However, diagnosing HIV infection requires two positive ELISA tests and one confirmatory test as positive well.

ELISA

These tests detect minimal amounts of antibodies, so small amounts of similar substances could lead to false positive results (vaccine against influenza, hepatitis B, and rabies), which are also caused by stimulation of the immune system (viral infections, pregnancy, autoimmune diseases, blood transfusion, *etc*.). It generally requires a Western blot test to confirm a positive result. The reaction is carried out in a plastic plate with microtiter wells. HIV antibodies and antigens cover the surface of each cavity. The patient's serum or plasma is then added to the well, and if the serum has specific antibodies, these bind to the plaque antigen complex. A labeled secondary antibody is then added to bind to the complex between the antibodies in the serum and the antigens on the plate. Finally, a chromogenic substrate reacts with the enzyme-labeled antibody. This union results in a change of color measured in the spectrophotometer. The optical density correlates with the antibody concentration of HIV in the patient: the higher the intensity, the more antibodies in the serum.

Rapid Tests

Rapid tests are anti-HIV antibody detection methods with results in less than 30 minutes, with a sensitivity and specificity range from 99.6 to 100%. In addition to plasma and blood serum, rapid tests use whole or capillary blood taken from the finger or ear lobe; other assays also use urine or oral transudation. The rapid tests are based on immunochromatographic methods, although the techniques use agglutination, immune dot blot, and immune filtration. They often use synthetic peptides or recombinant HIV proteins as a source of antigen and consist of a solid phase composed of particulates containing HIV antibodies and antigens. They are methodologically less complex than ELISAs combined with other tests for detecting antibodies; they constitute a strategy that improves the specificity of the results in an affordable way.

Western Blot

The technique consists of separating the viral proteins (antigens) by their molecular weight through electrophoresis, then transferring them to a nitrocellulose membrane to be used as a test strip. This strip is incubated with the patient's serum or plasma; if the serum has anti-HIV antibodies, it will bind to the antigen presented on the test strip. The resulting complex (antigen-antibody) is made visible using a secondary antibody and its corresponding substrate. The bands on the test strip are consistent with the antibodies in the sample. The following are the proteins that can be detected with this test, which correspond to the gene's structures of HIV: Envelope proteins: gp160, gp41, gp120, Polymerase proteins p31/p34, p39/p40, p51/p52, p66/p68, and Nucleocapsid proteins p17/p18, p24/p25, p55.

Follow-up Tests

During the clinical latency period associated with HIV infection, the virus continues to replicate. This period ends with the development of symptoms of the disease. The knowledge of the factors that affect the progression of the disease helps the comprehensive management of infection by HIV and the relationship between a high amount of HIV RNA and a decreased number of CD4+ T lymphocytes with rapid disease progression. The quantification of the viral load and the count of the CD4+ T cells provide prognostic information essential for the physician. A significant and progressive increase in viral load in the patient indicates treatment failure, but not if it is due to a lack of adherence to treatment or resistance to antiretrovirals.

Quantification of CD4+ T Lymphocytes

CD4+ T lymphocytes are essential for developing a specific immune response to infections, mainly intracellular pathogens. Since these cells are the target of HIV, their decline severely limits the host's immune response capacity. The ability of the immune system to have a specific response against HIV is a key factor in the course of the disease, and the quantification of CD4+ cells is a significant predictor of disease progression. Low CD4+ T lymphocyte counts are associated with an increased risk of disease progression. The recommendation to start treatment is based on the count of CD4+ T lymphocytes rather than any other marker. The risk of progression to AIDS is substantially increased when the CD4+ T lymphocyte count is <350 cells/mm^3. The risk is most significant when counts fall below 200 cells/mm^3. With this count, starting the antiretroviral treatment is recommended. The count of CD4+ lymphocytes is carried out by flow cytometry, using fluoresceine antibodies, and counts the percentage of CD4+ cells.

HIV RNA

The quantification of the HIV viral load is used as a prognostic marker to measure the progression of the disease and the efficacy of antiretroviral treatment. As mentioned above, elevated levels of viral load are associated with a decrease in CD4+ T lymphocytes, leading to progression to AIDS and death. Virological failure is the viral load with undetectable levels that is not achieved after six months of treatment (<50 copies/ml) or when you have a constant viral load of >50 copies/ml. Five assays to quantify HIV are currently approved by the FDA; two are RT-PCR tests, which are more sensitive and less susceptible to contamination, and the other is nucleic acid sequence-based amplification (NASBA). These tests are more than 98% sensitive to detect group M of HIV-1, except for real-time RT-PCR, which detects groups M, N, and O from HIV-1 [5].

Antiviral Therapy

Attempts have been made since the discovery of HIV to understand the pathophysiological mechanisms involved in the biological cycle of HIV, from virus interaction with the cell to viral particle release. With this information, the critical steps have been identified, and they are now the therapeutic targets for stopping the viral replication cycle. Antiretroviral treatment is the best option to achieve viral suppression for prolonged periods, keeping it at undetectable levels, thus increasing the CD4+ lymphocytes for as long as possible and consequently reducing mortality and morbidity; however, current treatments do not eradicate the infection of HIV.

Initially, the inhibition of HIV replication was directed to the viral enzymes, which are not present in human cells and are unique to the virus. Current treatments that inhibit HIV replication include entry of the virus into the cell and blocking the function of the three viral enzymes. Highly active antiretroviral treatment has shown to be effective in delaying the progression of AIDS by controlling viral replication for prolonged periods.

GLOSSARY

Acquired Immunodeficiency Syndrome (AIDS): A deadly disease that affects the human immune system and is caused by HIV infection.

Retrovirus: A virus replicates by reverse transcription of its RNA genome into DNA.

T-cells: A type of immune cell that helps the body fight infections and diseases.

Macrophages: A type of immune cell that engulfs and digests foreign substances, microbes, and cellular debris.

Dendritic cells: A type of immune cell that helps to activate T-cells and initiate an immune response.

Viral Envelope Protein: A protein that surrounds the genetic material of a virus and is responsible for its attachment to host cells.

gp120: The viral envelope protein of HIV that interacts with the host cell receptor CD4 and co-receptor CCR5 or CXCR4.

Host Cell Receptor: A protein on the surface of a host cell that is recognized by a virus and facilitates its entry.

CD4: The primary host cell receptor for HIV that is found on the surface of T-cells and other immune cells.

REVIEW QUESTIONS

1. What is the role of the viral envelope protein gp120 in HIV infection?

2. How does HIV evade the immune system's surveillance?

3. What is the global impact of HIV/AIDS on human health?

REFERENCES

[1] Menéndez-Arias L, Álvarez M. Antiretroviral therapy and drug resistance in human immunodeficiency virus type 2 infection. Antiviral Res 2014; 102: 70-86. [http://dx.doi.org/10.1016/j.antiviral.2013.12.001] [PMID: 24345729]

[2] Blanco JL, Varghese V, Rhee SY, Gatell JM, Shafer RW. HIV-1 integrase inhibitor resistance and its clinical implications. J Infect Dis 2011; 203(9): 1204-14. [http://dx.doi.org/10.1093/infdis/jir025] [PMID: 21459813]

[3] Briz V, Poveda E, Soriano V. HIV entry inhibitors: Mechanisms of action and resistance pathways. J Antimicrob Chemother 2006; 57(4): 619-27. [http://dx.doi.org/10.1093/jac/dkl027] [PMID: 16464888]

[4] Dwyer DE. Advances in HIV laboratory testing. N S W Public Health Bull 2010; 21(4): 86-8. [http://dx.doi.org/10.1071/NB10018] [PMID: 20513308]

[5] Friedrich BM, Dziuba N, Li G, Endsley MA, Murray JL, Ferguson MR. Host factors mediating HIV-1 replication. Rev 2011; 161(2): 101-14. [PMID: 21871504]

[6] Peeters M. The genetic variability of HIV-1 and its implications. Transfus Clin Biol 2001; 8(3): 222-5. [http://dx.doi.org/10.1016/S1246-7820(01)00131-8] [PMID: 11499964]

CHAPTER 18

Molecular Mechanisms of Human Immunodeficiency Virus Resistance to Antiretroviral

Saima Pervaiz[1,*] and **Shehryar Awan**[2]

[1] *Department of Immunology, University of Health Sciences, Lahore, Pakistan*

[2] *Department of Pathology, Azra Naheed Dental College, Superior University, Lahore, Pakistan*

Abstract: Antiretroviral therapy (ART) has transformed the treatment of human immunodeficiency virus (HIV) infection, improving life expectancy and quality of life for millions worldwide. However, the emergence of drug-resistant HIV strains poses a significant challenge to the effectiveness of ART. The molecular mechanisms underlying HIV resistance to antiretroviral drugs involve multiple genetic changes in the viral genome that reduce drug susceptibility, often through alterations in the viral enzymes targeted by the drugs. The primary targets of ART are the viral reverse transcriptase (RT), protease (PR), and integrase (IN) enzymes, which are essential for HIV replication. Resistance to nucleoside reverse transcriptase inhibitors (NRTIs) results from mutations in the viral RT enzyme that reduce drug incorporation into the viral DNA chain. Non-nucleoside reverse transcriptase inhibitors (NNRTIs) bind to a hydrophobic pocket near the RT active site, and resistance to these drugs arises from mutations that alter the binding pocket conformation. Protease inhibitors (PIs) bind to the viral PR enzyme, and resistance results from mutations that alter the enzyme's conformation, reducing drug binding affinity. Integrase strand transfer inhibitors (INSTIs) bind to the viral IN enzyme, and resistance arises from mutations that affect drug binding or alter the IN active site. The emergence of drug-resistant HIV strains can also result from poor adherence to ART, leading to the selection of pre-existing resistant viruses or the development of new resistance mutations. In addition, the genetic diversity of HIV and the high viral replication and mutation rate contribute to the rapid evolution and emergence of drug-resistant strains.

Keywords: Antiretroviral therapy, Drug resistance, Human immunodeficiency virus, Nucleoside reverse transcriptase inhibitors (NRTIs), Integrase, Non-nucleoside reverse transcriptase inhibitors (NNRTIs), Protease, Protease inhibitors, Viral genome.

* **Corresponding author Saima Pervaiz:** Department of Immunology, University of Health Sciences, Lahore, Pakistan; E-mail: shehryarawan73@gmail.com

Sanaullah Sajid, Sajjad ur Rahman, Shahid Mahmood, Shayan Bashir & Mudasser Habib (Eds.)

INTRODUCTION

In the last two decades, there have been notable advancements in the management of human immunodeficiency virus (HIV) infection, resulting in improved life expectancy and enhanced quality of life for patients. These improvements can be attributed primarily to the expanding repertoire of Antiretrovirals (ARVs) designed for HIV treatment, which encompasses diverse molecular approaches. Despite the notable efficacy of Highly Active Antiretroviral Therapy (HAART), a treatment regimen that involves the administration of a combination of three or more medications with distinct modes of action, it is important to acknowledge that the total suppression of viral replication or complete elimination of HIV from the reservoir within the host has not been achieved. It is only feasible to refer to an efficacious control of viral replication for prolonged periods.

Failure to adhere to the therapeutic regimen after the first year of initiating treatment is expected in 10 to 30% of individuals on initial therapy and between 40 and 60% of patients who have previously failed treatment. The leading causes of antiretroviral treatment failure are resistance to ARVs, limited antiviral potency, inadequate patient adherence to the therapeutic regimen, defects in drug absorption, drug interactions, insufficient drug activation (cellular resistance), and, finally, viral replication in lymphoid sanctuaries. ARV resistance is the leading cause of treatment failure. It is due to a series of nucleotide substitutions in the genes coding for glycoproteins 120 and 41, the reverse transcriptase (RT) enzymes, protease, and integrase, targets on which these drugs act [1].

ANTIRETROVIRAL TREATMENT

Since the discovery of HIV, therapeutic strategies have focused on the virus's chemical structure and life cycle, which led to the creation of the drugs available today. Thus, as the understanding of the pathophysiological mechanisms involved, the interaction of the viral envelope with specific receptors in the cell membrane, replication, and even the release of new infectious particles, new options for inhibiting them emerge. In the virus cycle, critical steps have been identified and targeted by ARVs currently used to inhibit replication, such as entry, reverse transcriptase, protease, and integrase. ARV treatment offers clear benefits to HIV-infected individuals when HAART is used, which usually involves the combination of two nucleoside analogues and a protease inhibitor, improves the quality and quantity of the patient's life, as well as decreases the frequency of some infections by opportunistic microorganisms.

At present, the goal of treatment is to maintain the viral load at undetectable levels and to increase the CD4+ lymphocyte count progressively for as long as possible. There are more than 20 different ARVs for the treatment of HIV patients.

THERAPEUTIC TARGETS [2]

Envelope Proteins

HIV attaches to the cell through the interaction of the virus envelope glycoprotein complex (gp160) with the CD4+ lymphocyte surface receptor and, depending on the course of the infection, with the co-receptors CCR5 and CXCR4.15. The glycoprotein gp160 is composed of two subunits, the glycoprotein gp120, which is the one that interacts with cellular receptors, and the transmembrane subunit gp41, which is the cause of virus fusion. After the binding of gp120 and the CD4+ cell receptor, a conformational change occurs in the gp120 that allows a closer interaction between the co-receptors CCR5 and CXCR4 on the surface of the CD4+ cell and gp120, causing a change in gp41, which results in the exposure of the hydrophobic region of the central part of the gp41 protein. Close to this region are the HR1 and HR2, composed of six helices forming a "hairpin" structure and, consequently, the fusion pore, through which the viral RNA enters the CD4+ lymphocyte.

Cell Receptor Inhibitors

CCR5 and CXCR4 inhibitors are the most promising within this class of drugs since blocking the receptor achieves a similar effect to that seen in people who possess a mutation in the CCR5 allele and generate a truncated, non-functional protein. Thus, when homozygotes do not express CCR5 and have relative protection against HIV infection, heterozygotes, on the other hand, can acquire the infection, but disease progression is slower. On the other hand, CXCR4 is found in 50% of patients with advanced disease, and its presence in lymphocytes is associated with rapid progression. Since CXCR4 is expressed with its ligand SDF-1α (stromal cell-derived factor-1α), when SDF-1α is co-inhibited with CXCR4, hematopoiesis, cardiac, and brain development functions are altered parallelly.

Among the inhibitors of CCR5, we can find new drugs such as Maraviroc, a non-competitive inhibitor of the co-receptor binding site, and Vicriviroc, which, like TAK-770, induces conformational changes in CCR5 that affect gp120 binding. Among the CXCR4 inhibitors, AMD070 blocks SDF1α, with the side effects described in the preceding paragraph.

Fusion Inhibitor

At present, there is only one inhibitor of this class, Enfuvirtide. This peptide of 36 amino acids prevents the termination of the HIV fusion sequence. Enfuvirtide is a synthetic analogue of the peptide fragment HR2, which binds to an HR1 moiety and destabilizes the protein hairpin formation required for complex pore formation and fusion of the virus to the cell membrane.

Protease Structure and Function

HIV protease is a symmetrical dimer with 99 amino acids, with an active site in the cleft at the interphase between the two monomers from position 25 to 27 of each monomer. The function of this enzyme is to cleave the polyprotein formed as a product of the translation of the mRNA of the gag, pol, and env genes to transform them into functional or active proteins of the internal capsid (p17, p24, p7, and p6), essential enzymes (RT, protease, and integrase), and envelope glycoproteins (gp120 and gp41), which are required for the production of mature, infectious viruses.

Protease Inhibitors

Currently used protease inhibitors bind to the active site by two mechanisms: the first simulates the transition state during peptide cleavage, and the second binds to the active site as a steric complement. When protease function is blocked, the formation of the internal capsid proteins is disrupted, causing the assembly of immature virions not to take place, and therefore they lose their infectivity. Nine inhibitors of this enzyme are currently approved for clinical use (Table **1**).

Table 1. Antiretroviral drugs are used in the treatment of HIV-infected individuals.

Class	Antiretrovirals	Abbreviation	Target	Mechanism of Action
Entry inhibitor	Vicriviroc Maraviroc AMD070	SCH 471690 UK-427,857	CCR5 CCR5 CXCR4	Induces conformational changes in CCR5 affecting gp120 binding. Blocks SDF1α
Inhibitor of reverse transcriptase analogue of the nucleosides Nucleotide analogue	Zidovudine Didanosina Zalcitabine Estavudina Lamivudine Abacavir Gemcitabine Tenofovir	AZT Ddl DdC D4T 3TC ABV FTC TFV	Reverse transcriptase	Phosphorylated by cellular enzymes. Competitively inhibit viral DNA synthesis or cause viral chain termination.
Reverse transcriptase inhibitor non-nucleoside analogue	Nevirapine Delavirdine Efavirenz	NVP DLV EFV	Reverse transcriptase	Non-phosphorylated, they bind directly to the enzyme and prevent transcription.

(Table 1) cont.....

Class	Antiretrovirals	Abbreviation	Target	Mechanism of Action
Protease inhibitors	Saquinavir Indinavir Ritonavir Nelfinavir Amprenavir Lopinavir Atazanavir Fosamprenavir tipranavir Darunavir	SQV IDV RTV NFV APV LPVr ATV FAPV TPV DAR	Protease	Bind to the enzyme's active site inhibits proteins' proper assembly and conformation.
Integrase inhibitors	Elvitegravir Raltegravir	GS 9137 MK-0518	Integrase	Inhibit the integration process of the viral pre-integration complex to cellular DNA.
Fusion inhibitors	Enfuvirtide	T-20	Wrap gp41	Bind to the HR1 region of gp41 and prevent fusion of the virus with the CD4+ cell.

STRUCTURE AND FUNCTION OF REVERSE TRANSCRIPTASE

The process of reverse transcription involves the conversion of viral RNA into double-stranded DNA by the enzyme reverse transcriptase (RT), which subsequently integrates this DNA into the genome of the host cell. The heterodimeric enzyme, consisting of a 66 kDa protein (p66) and a 51 kDa protein (p55), is generated through the protease-mediated processing of the Gag-pol precursor. The p66 protein can be subdivided into two distinct subunits, namely p51 and p15. The p51 segment is associated with polymerase activity, while p15 is associated with RNase activity. The optimal polymerase activity necessitates the presence of both proteins, namely p66 and p15 [3].

The enzyme's structure was determined by crystallography, and some of its properties were deduced. A schematic representation of p66 can be made by comparing its structure and function to a right-handed with structural features called "palm", "thumb", and "fingers". The initiating template passes through the cleft between the fingers and thumb, where the enzyme's catalytic site is located. The deoxynucleotide (dNTP) incorporation occurs in the first two fingers, and the growth of the temperate is processed through the palm of the hand, conserved (MDD motif sequence) comprising the amino acids encoded by codons 183 to 186. The *in vitro* polymerase activity is diminished with substitution of any of these amino acids. This particular activity is linked to the process of substrate binding, involving the interaction between the complex and specific amino acid regions, namely amino acids 73-90 and 141-174 with the template, and amino acids 227-235 with the initiator. The interaction between dNTPs and amino acids

65, 72, 110, 113, 115, 116, 151, 160, 183, 184, 186, and 219 is responsible for the binding process, which is facilitated by the presence of magnesium.

Reverse Transcriptase Inhibitors

The first ARV approved in 1987 for the treatment of HIV was zidovudine which belongs to this class of inhibitors. Its success led to the emergence of other inhibitors, both analogous and non-nucleoside analogue inhibitors. There are currently 11 ARVs that act by blocking this enzyme: eight are nucleoside analogues, and three are non-nucleoside analogues.

Reverse Transcriptase Inhibitors Nucleoside Analogues

The nucleoside analogue reverses transcriptase inhibitors (NRTIs) have a similar molecular structure to natural deoxyribonucleoside triphosphates (dNTPs) found in the nucleus and used in the process of DNA synthesis from HIV RNA. NRTIs prevent this process in two ways: first, by preventing dNTPs from being used, and second, by causing the DNA strand to end prematurely due to a change in the 3′ hydroxyl group. Due to its absence, the phosphodiester bond between the expanding DNA strand and the new 5′ triphosphate nucleoside is not formed in the presence of these ARVs.

Reverse Transcriptase Inhibitors Non-Nucleoside Analogues

Non-nucleoside analogue reverse transcriptase inhibitors (NNRTIs) bind to the hydrophobic gap in the palm near the enzyme's active site, causing a conformational change. This binding inhibits RT function by an allosteric mechanism whereby the asparagine is displaced to the polymerase binding site.

LEADING CAUSES OF FAILURE TO ANTIRETROVIRAL TREATMENT

Despite the success of HAART in treating HIV patients, many patients experience treatment failure; it is estimated that almost 30% of individuals fail to suppress or decrease viral RNA to undetectable levels, despite an effective therapeutic combination. In addition, 40-60% of subjects, after one year of HAART initiation, experience viral "rebound" after viral RNA has been suppressed to undetectable levels. Factors contributing to therapeutic failure are:

a) Inadequate adherence to ARVs, due to complex pharmacological schemes, tolerance, schedules, other concomitant medications, *etc.*; lack of adherence results in reduced treatment potency.

b) Pharmacological factors, such as drug-drug interactions and poor gastrointestinal absorption of ARVs.

c) Host factors, inadequate activation of ARVs (NRTI), viral load, and CD4+ lymphocyte count at the initiation treatment.

d) Limited potency of some HAART regimens.

e) Viral replication in "sanctuaries".

f) Resistance to ARVs.

ARVs do not directly cause the mutations that lead to resistance; they alter the intracellular environment since the mutations are spontaneous; if these mutations confer an advantage over non-resistant viruses, the population of mutated viruses will grow faster than the "normal" or natural type and become the dominant population in the patient. Since these viruses with resistance mutations can replicate in the presence of ARVs, they cause the viral load to increase and disease progression.

FACTORS INVOLVED IN RESISTANCE TO ANTIRETROVIRALS

ARV resistance is the leading cause of treatment failure and is generated by a series of nucleotide substitutions in the genes coding for the gp41 and for the RT and protease enzymes, targets in which ARVs work. The reverse enzyme transcriptase drives the rate of mutation of the HIV genome; it lacks exonuclease activity in the incorporation of bases during double-stranded DNA synthesis, where another can randomly substitute a nucleotide. The HIV genome contains approximately 10,000 bases, and the error rate of RT is estimated to be about one nucleotide per 10,000 nucleotides copied, equivalent to one point mutation in each replication cycle.

The daily production of 1 to 10 trillion HIV particles, coupled with the daily replacement of more than half of the viral population every day and the selective pressure of both drugs and the immune system, favour the induction of resistance to ARVs. Most variants generated by HIV mutations have little or no effect on viral function or replication capacity, and many others may be lethal to the virus. However, mutations at important sites, *e.g.*, in viral enzyme genes, affect the virus's ability or may confer resistance to one or more ARVs.

Another factor contributing to the genetic diversity and resistance of ARVs is the recombination of the two RNA strands. This phenomenon is an inherent component of the typical viral replication cycle and potentially facilitates the restoration of faulty viral genomes, which enhances viral variation or expedites the dissemination of advantageous mutations among distinct viral subgroups.

Antiretroviral Resistance Mutations

The development of ARV resistance is the biggest obstacle to the success of HAART. Resistance emerges at different rates in different ARVs. When resistance is assessed at first virologic failure, it is usually associated with ARVs that are "weaker" or more likely to generate resistance on their own rather than all the components of the therapeutic regimen. The genetic barriers of the ARV itself explain the difference in the emergence of resistance to a specific treatment component. For NNRTIs and lamivudine (3TC), a single mutation can confer resistance and, in some cases, generate resistance for all those in the same class. Developing high resistance levels requires several mutations for the rest of the NRTIs and protease inhibitors. Continuous replication in the presence of ARVs leads to the accumulation of several mutations, ultimately allowing cross-resistance development within the same ARV class and affecting future efforts to reduce viral replication again. A genetic barrier is said to be present when multiple mutations are necessary to confer resistance.

Resistance to Cell Receptor Inhibitors

The blockers of these receptors act on the cell instead of some viral protein. Some ways of acquiring resistance are:

a) In the presence of CCR5 inhibitors, the virus can switch from CCR5 - tropic (R5) to CXCR4 - tropic (X4).

b) The virus can evolve to have a higher co-receptor affinity, use low levels of co-receptor for replication and pathogenesis, and the natural variability of the exposed region of the gp120.

Fusion Inhibitor Resistance

Resistance to this ARV is found near the N-terminal HR-1 region of gp41, which is highly conserved and indispensable for the fusion process. A single mutation at positions 36 and 38 in the HR1 domain significantly decreases susceptibility to Enfuvirtide. The amino acid changes from HR1 and gp120 are believed to be associated with the virus's susceptibility to Enfuvirtide. This explains the wide range of susceptibility among HIV strains to this inhibitor.

Resistance to Protease Inhibitors

Amino acid substitutions that lead to protease resistance are located near the enzyme's active site or in amino acids outside the active region. These amino acid changes directly or indirectly modify the number and nature of the contact points between the inhibitors and the protease; thus, the enzyme's affinity is reduced. Mutations in protease are classified into primary or major mutations and secondary or minor mutations. Primary mutations are those that reduce the susceptibility to ARVs per se. In contrast, secondary mutations decrease susceptibility to ARVs by themselves, while secondary mutations decrease the susceptibility or enhance the replication capacity of strains with primary mutations. Some mutations may be considered preliminary for one ARV but secondary for another, so they cannot be generalized to all ARVs or drug groups.

Generally, high resistance levels require a series of amino acid substitutions, which usually vary between protease inhibitors. Some mutations are selected for only one inhibitor, reflecting particularities in the chemical structure of the inhibitor. However, many overlapping combinations of mutations develop resistance to these inhibitors, explaining the wide range of cross-resistance. One example of a primary resistance mutation is D30N which confers resistance to nelfinavir (NFV) only. The mutations characteristically confer cross-resistance among protease inhibitors which include codons 82, 84, and 90. Another example is codon I50V which causes resistance to amprenavir (Apv) and contributes cross-resistance to lopinavir (LPVr).

Resistance Inhibitors to Reverse Transcriptase Nucleoside Analogues

All resistance mutations appear to influence the dNTP binding site and the enzyme's binding site to the initiator template. These mutations are thought to increase enzyme specificity in recognizing natural dNTPs. Several mutations are required in RT to show high phenotypic resistance to these inhibitors. Nevertheless, the frequency of mutations varies for each antiretroviral (ARV), with the exception of 3TC, which confers resistance by a solitary mutation at codon 184. Every nucleoside reverse transcriptase inhibitor (NRTI) elicits a consistent set of changes. For instance, the acquisition of resistance to AZT occurs through the successive emergence of certain mutations, namely codons 41, 67, 70, 210, 215, and 219, which are commonly referred to as thymidine analogue mutations (TAMs).

Resistance to didanosine (ddI) is characterised by the presence of specific codons, namely 65, 74, and 184. On the other hand, resistance to zalcitabine (ddC) is associated with the presence of codons 65, 69, 74, and 184. Nevertheless, it is worth noting that there is a phenomenon of mutation overlap observed among many antiretroviral drugs (ARVs) belonging to the same class. This overlap ultimately results in the development of cross-resistance, wherein resistance to one drug within the class confers resistance to other drugs in the same class. Various mutations within the RT gene give rise to diverse levels of cross-resistance. In certain instances, these changes result in resistance to a particular set of antiretroviral drugs (ARVs), whilst in other cases, resistance to all ARVs within the group occurs (Table **2**). One instance of a mutation, namely the 65R mutation, has been observed to undergo selection and result in a varied reduction in susceptibility to ddI, abacavir (ABV), and tenofovir (TFV). According to prior research, this mutation is highly probable to result in cross-resistance to all IRTANs, with the exception of AZT. There are three types of multi-nucleoside resistance that have been identified: the thymidine analogue mutations (TAMs), the insertion complex at codon 69, and the Q151M complex. The binding of all non-nucleoside analogue reverse transcriptase inhibitors to a common hydrophobic gap elucidates the observed cross-resistance within this inhibitor class. Specifically, the presence of either the K103N or Y188L mutation alone significantly impacts the clinical efficacy of these medications.

Table 2. Types of resistance to antiretrovirals.

Antiretrovirals	Mutations
NRTIs	
Abacavir	K65R, L74V, Y115F, M184V
Didanosine	K65R, L74V
Emtricitabine	K65R, M184VI
Lamivudine	K65R, M184VI
Stavudine	M41L, D67N, K70R, L210W, T215W, T215YF, K219QE
Tenofovir	K65R, K70R
Zidovudine	M41L, D67N, K70R, L210W, T215YF, K219QE
NNRTIs	
Delavirdine	K103N, V106M, Y181C, Y188L, P236L
Efavirenz	L100I, K103N, V106M, V108I, Y181CI, Y188L, G190SA, P225H
Nevirapine	L100I, K103N, V106AM, 108I, Y181CI, Y188CI, Y188CLH, G190A
Multi-Resistance	
INSERTION 69	M41L, A62V, 69, K70R, L210W, T215YF, K219QE
COMPLEX 151	A62V, V75I, F77L, F116Y, Q151M
TAM	M41L, D67N, K70R, L210W, T215YF, K219QE

(Table 2) cont.....

Antiretrovirals	Mutations
	IP
Atazanavir ± RitonavirFosamprenavir/Ritonavir	L10IFVC, G16E, K20RMITV, L24I, V32I, L33IFV, E34Q, M36ILV, M46IL, G48V, I50L, F53LY, I54LVMTA, D60E, I62V, I64LMV, A71VITL, G73CSTA, V82ATFI, I84V, I85V, N88S, L90M, I93LML10FIRV, V32I, M46IL, I47V, I50V,
Darunavir/Ritonavir Indinavir/Ritonavir Lopinavir/Ritonavir Nelfinavir Saquinavir/Ritonavir Tipranavir/Ritonavir	I54LVM, G73S, V82AFST, I84V, L90M V11I, V32I, L33F, I47V, I50V, I54ML, G73SA, L75V, I84V, L89V L10IRV, K20MR, L24I, V32I, M36I, M46IL, I54V, A71VT, G73SA, V77I, V82AFT, I84V, L90M L10FIRV, K20MR, L24I, V32I, L33F, M46IL, I47VA, I50V, F59L, I54VLAMTS, L63P, A71VT, G73S, V82AFTS, I84V, L90M L10FI, D30N, M36I, M46IL, A71VT, V77I, V82AFTS, I84V, N88DS, L90M L10IRV, L24, G48V, I54VL, I62V, A71VT, G73S, V77I, V82AFTS, I84V, L90M L10V, I13V, K20MR, L33F, E35G, M36I, K43T, M46L, I47V, I54AMV, Q58E, H69K, T74P, V82LT, N83D, I84V, L90M
	IF
Enfuvirtide	G36DS, I37V, V38AME, Q39R, Q40H, N42T, N43D

*NRTIs, nucleoside analogue reverse transcriptase inhibitors; NNRTIs, non-nucleoside analogue reverse transcriptase inhibitors; TAM, thymidine analogue mutations; PI, protease inhibitors; FI, fusion inhibitors.

Resistance to Integrase Inhibitors

Resistance to these inhibitors is related to mutations at the enzyme's catalytic site that causes conformational changes. The degree of resistance depends on the number of mutations in the integrase. Some of the mutations have more effect on the viability of the viral variants, *e.g.*, the C130S mutation, which impairs viral replication. It is considered to be an inhibitor with a low genetic barrier. The resistance profile of Elvitegravir has been observed *in vitro*, selecting mutations H51Y, E92Q, S147G, and E157Q; also been linked to the T66I and R263K. In the case of Raltegravir, resistance mutations have not been disclosed, but it has been noted that the E92Q mutation reduces susceptibility, generating cross-resistance with Elvitegravir.

HIV Subtypes and Resistance to Antiretrovirals

Current knowledge about the development of ARV resistance in HIV-1 is based on studies of subtype B strains, the predominant subtype in developed countries. However, the introduction of ARVs in other parts showed that the viral subtype influences treatment efficiency. HIV-2 and HIV-1 subtype O have the amino acid isoleucine and cysteine in their natural consensus at codon 181, respectively, naturally generating resistance to all NNRTIs. In the absence of ARV treatment, it was found that the RT and protease of HIV-1 of subtype B and non-B are different by 40 and 30% of amino acids, respectively. In viruses with non-B subtypes, some of these substitutions are naturally present and occur with high frequency in the positions associated with resistance in subtype B.

TESTS FOR THE DETECTION OF RESISTANCE

ARV resistance testing has been rapidly incorporated into comprehensive patient management. They help determine which ARVs are not working, and which can be more efficient or used longer. There are two tests to assess ARV resistance: phenotypic and genotypic [4].

Phenotypic Tests

This type of test measures the ability of the virus to replicate in the presence of various concentrations of ARVs. These tests provide a direct measure of the sensitivity of the drug, with results generated as an inhibitory index at 50 or 90% (IC50 or IC90) and the amount of ARV that inhibits replication at 50 or 90% (IC50 or IC90). Another advantage is that they help provide information on cross-resistance, and these tests are easy to interpret for treating physicians. Among the disadvantages of phenotypic tests compared to genotypic tests are: that they are very laborious, much more expensive, are performed in a limited number of laboratories, and the processing time and delivery of results are longer.

The cut-off values are established and standardized for only a few ARVs; furthermore, viral populations below 10-25% and early mutations are not detected. In some phenotypic tests that do not use recombinant technology but instead use peripheral blood mononuclear cells, some variants can adapt or change rapidly in the culture medium but without being the predominant ones in the patient, giving an erroneous resistance result.

Genotypic Tests

They are used to detect specific mutations in the protease gene and RT associated with phenotypic resistance to a particular ARV. Three types of methodologies are available:

- Hybridization of the amplified product.
- The use of “chips” examines multiple codons by hybridization.

• Sequencing analysis of the pol gene.

These tests detect mutations in populations above 5 to 20%; they are simpler to perform compared to phenotypic tests; the time to obtain results is much shorter, and they can detect early mutations before the appearance of phenotypic resistance. Disadvantages of genotypic tests include; indirectly measuring resistance, interpretation of the results is more difficult, as experience in sequence analysis is required, and the use of logarithms that correlate resistance mutations with data on known mutations that confer phenotypic resistance. Finally, for some mutations, their clinical significance is not known.

ARV resistance tests, commonly known as genotypes, require controlled conditions from the time of application to the technique's performance. Patients should not discontinue treatment for at least two weeks before blood sampling. The natural strain has a higher replication capacity than strains with resistance; therefore, with discontinuation of the treatment, the genotype will not detect the resistance strains, which will reappear when the patient initiates treatment. Preferably, the levels of plasma viral RNA must be greater than 5,000 copies to detect circulating resistance; it is also desirable that the plasma separation is done immediately after extraction and in a sterile environment and with RNase-free material to avoid degradation. The blood sample should be processed in certified laboratories with strict quality control. The technique consists of extracting viral RNA by density gradient and solvents, then amplifying the virus's protease and RT regions. Finally, the purification and sequencing of the products were obtained. The latter results are analyzed in a constantly updated database with the internationally known resistance codons for each of the ARVs. An expert must perform the verification to eliminate errors that occur during the routine procedure. The shipment of samples several hours between collection and processing or using filter paper leads to false results that must be considered in their clinical application.

Clinical Application

There is sufficient data on the benefits and limitations of ARV resistance testing in selecting the most efficient therapeutic regimen. However, there are specific considerations concerning treatment failure, switching, and resistance testing. Resistance tests do not define treatment failure; the increase of viral RNA levels, the drop in viral lymphocyte counts and CD4+ lymphocyte count, and clinical progression should be considered for bringing a change in the treatment. In addition, the history of previous treatments should be considered while choosing a new treatment method and the use of ARV or ARVs with proven resistance should be avoided.

The International AIDS Association of the United States, the European Guidelines, and the Medical Health Care Guidelines recommend the use of these tests under the following circumstances:

a) Individuals without previous ARV treatment with primary infection or with less than 12 months of seroconversion for the detection of transmission of resistant strains.

b) Suboptimal response to therapy; when the viral load is not reduced in eight weeks, it suggests the pre-existence of resistance.

c) Before initiating treatment in patients with established infection; when the regional prevalence of resistance is greater than 5%, although in some cases, the wild strain is detected since the resistant strain remains a minority.

d) After the first treatment failure.

e) Change of therapy; Only if it is suspected to be due to resistance, with viral loads above 500 to 1,000 copies.

f) Multiple therapeutic failures.

g) Pregnancy. If viral RNA is detected in the mother as a prophylaxis in vertical transmission.

PREVALENCE AND IMPACT OF RESISTANCE TO ANTIRETRO-VIRALS

Patients without Previous Antiretroviral Treatment

The detection of mutations related to ARV resistance in individuals not exposed to ARVs is primarily due to the transmission of resistant strains from previously treated subjects or by spontaneous mutations of the virus. The prevalence of resistance in previously untreated patients observed in Europe and the United States was above 10 and 20%, respectively. Latin America reports similar prevalence in individuals with recent infection; with Argentina showing 15.6%, and Brazil 6.6% prevalence of resistance.

There are few studies on the prevalence of resistance to ARV in previously untreated patients with established HIV infection; in the United States, Canada, and Mexico, a prevalence between 8.3 and 20% was found, while Europe and Belgium reported a prevalence of 7.2%. The transmission of resistance has increased in recent years, resulting in a poor response to ARV treatment in previously untreated patients, making it a serious health problem.

Patients with Previous Antiretroviral Treatment

For patients with prior antiretroviral treatment, the prevalence of resistance in infected individuals with detectable viral RNA is approximately 78% in the United States and Canada. The NRTIs represent the highest resistance at 70%, followed by protease inhibitors with 42% and 31% for NNRTIs; patients who develop the most resistance include those with the most advanced disease. This high prevalence has implications for the transmission of resistant strains and efficient intervention of ARV treatment.

ECONOMIC IMPACT

Although this type of testing is costly, its rational use helps support treatment decisions that have positive cost-effectiveness, as the administration of drugs to which there is resistance is an unnecessary financial outlay. The expense caused when a patient experiences treatment failure is increased, as diseases caused by opportunistic microorganisms, hospitalization, and use of diagnostic services have increased; therefore, additional drugs are required for their control, similarly more staff is needed to care for the patient, and, finally, there is a decrease in patient productivity, increased disability and an impact on the national economy.

GLOSSARY

Acquired Immunodeficiency Syndrome (AIDS): A virus that infects humans and causes illness in the immune system. The loss of CD4+ cells is a hallmark of AIDS. Infections that could kill the host are made available in this way.

Adherence: The capacity to adhere to a particular therapeutic regimen. Medication adherence refers to consuming the prescribed prescription dosage at the appropriate time and proof of consuming the prescribed dosage at the appropriate time, with the provided instructions. Noncompliance with an HIV treatment regimen can result in treatment failure and the development of medication resistance.

Antiretroviral Therapy: Drug treatment that stops viruses from spreading in the body. Highly active antiretroviral therapy (HAART) is the antiretroviral therapy that is suggested for people with HIV.

CD4+ Cell: An infection-fighting type of white blood cell. When it comes to the immune system, the more CD4+ cells a person has, the better. HIV attacks CD4+ cells and kills them. This makes the immune system weaker.

CD4+ Cell Count: The quantification of CD4+ lymphocytes in a blood specimen. This tool assists healthcare practitioners in determining the appropriate initiation of HIV therapy and evaluating the efficacy of treatment interventions. The typical range for a CD4+ cell count is commonly observed to fall within the values of 500 to 1400.

Drug Resistance: When pathogens undergo adaptive changes that enable them to resist the effects of a pharmaceutical agent and persist in their replication. This phenomenon may occur even when an individual administers medications that typically combat the ailment.

First-Line Regimen: The initial pharmacological regimen and therapeutic interventions employed during the commencement of therapy for a medical condition.

Highly Active Antiretroviral Therapy (HAART): Treatment regimens aimed at inhibiting or decelerating the replication of the human immunodeficiency virus (HIV) and preventing the advancement of HIV-related illness. The standard Highly Active Antiretroviral Therapy (HAART) regimen typically involves the concurrent administration of three or more antiretroviral medicines targeting the Human Immunodeficiency Virus (HIV), selected from a minimum of two distinct therapeutic classes.

Human Immunodeficiency Virus (HIV): AIDS is caused by a virus. HIV-1 causes most HIV cases around the world.

Immunodeficiency: Not being able to make enough disease-fighting cells to keep the body safe from infections.

Immunosuppression: Not being able to fight off sickness as well as it should. Things like medication or diseases like HIV can bring it on.

REVIEW QUESTIONS

1. Why is it important to take antiretroviral medication regularly?
2. Why must certain drugs be taken with food and others on an empty stomach?
3. Discuss the challenges of ART.

REFERENCES

[1] Krambovitis E, Porichis F, Spandidos DA. HIV entry inhibitors: A new generation of antiretroviral drugs. Acta Pharmacol Sin 2005; 26(10): 1165-73. [http://dx.doi.org/10.1111/j.1745-7254.2005.00193.x] [PMID: 16174430]

[2] Clavel F, Hance AJ. HIV drug resistance. N Engl J Med 2004; 350(10): 1023-35. [http://dx.doi.org/10.1056/NEJMra025195] [PMID: 14999114]

[3] Viganò A, Schneider L, Giacomet V, *et al.* Efficacy and tolerability of multiple drug therapy in HIV-infected children. J Infect 2005; 50(5): 404-11. [http://dx.doi.org/10.1016/j.jinf.2004.08.001] [PMID: 15907548]

[4] Geretti AM, Easterbrook P. Antiretroviral resistance in clinical practice. Int J STD AIDS 2001; 12(3): 145-53.
[http://dx.doi.org/10.1258/0956462011916938] [PMID: 11231866]

[5] Kuritzkes DR. Extending antiretroviral therapy to resource-poor settings. AIDS 2004; 18 (Suppl. 3): S45-8.
[http://dx.doi.org/10.1097/00002030-200406003-00009] [PMID: 15322484]

SUBJECT INDEX

A

Sanaullah Sajid, Sajjad ur Rahman, Shahid Mahmood, Shayan Bashir & Mudasser Habib (Eds.)

D

E

K

L

M

N

O

P

R

X

Y